Laboratory Manual for Animal Technicians

VICTORIA SOLBERG

ILLUSTRATED BY **EDWIN RIVERA** AND **VICTORIA SOLBERG**

IOWA STATE UNIVERSITY PRESS / AMES, IOWA

VICTORIA SOLBERG is a medical manuals writer/editor for the United States Army Medical Bioengineering Research and Development Laboratory. She has taught laboratory animal science courses at Wilson College, as well as courses in parasitology, animal husbandry, and pharmacology. She holds the M.S. degree in biology from Fairleigh Dickinson University.

Printed by
The Iowa State University Press
Ames, Iowa 50010

First edition, 1985
International Standard Book Number:
0-8138-1066-3

Contents

Preface

This manual is designed for students who have had little or no experience working with laboratory animals or administering drugs or medication. It describes and illustrates techniques, procedures, and equipment and allows students to observe the actions of a variety of drugs on laboratory animals.

Most of the drugs and equipment used in the manual are normally found in a veterinary practice. Three lists are included at the end of the manual: two lists of commercial products (drugs and equipment) plus a list of suppliers and their addresses. These lists are not intended to be a recommendation of certain products to the exclusion of others, but the products listed are those the author is familiar with and has found to be satisfactory.

Every effort has been made to ensure that the dosage recommendations are precise and officially accepted at the time of publication. Several drugs have been used by the author as experimental drugs. Anyone using them should check the manufacturer's recommendations for dosage and use them with caution and discretion.

Directions are written for a right-handed technician. Left-handed individuals must use the opposite hand.

This manual was developed under a grant from the Department of Education. The contents, however, do not necessarily represent the policy of that agency, nor is there specific endorsement by the federal government.

Thanks are extended to the following individuals who helped make this manual possible: Jo Morisse, for initial encouragement and inspiration; Dr. Don Ruch, for his photography skills, technical advice, encouragement, and total support; Dr. A. A. Gruen, for his support and technical advice in veterinary medicine; Donna Harlacher, for her typing skills, patience, and good humor; Margaret Taylor, for her encouragement and editorial skills; Wendy Shelton, for her editorial skills; Edwin Rivera, for his assistance as an illustrator; and all other faculty members of Wilson College for their encouragement and support.

Laboratory Manual for Animal Technicians

EXPERIMENT 1

Oral Administration of Ethanol in the Mouse

INTRODUCTION

This experiment acquaints the student with the syringe, restraining and intubating (dosing orally) mice, and behavioral and/or physiological effects of ethanol in mice.

In this experiment, 10 mice are intubated with water, 10 mice are intubated with ethanol (ETOH), and 10 are untreated. Physiological and/or behavioral responses of the three groups are then compared to determine the effects of ethanol (the drug-treated group) and of handling (the water-treated group) in mice.

MATERIALS AND EQUIPMENT (per student)

- mouse scale and balance weights (per 2 students)
- 30 mice, each approximately 20 g (held without food 4 hr before experiment)
- syringe, 3 cc
- feeding needle, 18 gauge, 1 1/2 in.
- beaker with 25 cc tap water
- beaker with 25 cc 35% ethanol (ETOH)
- grid (optional)
- 2 empty cages

PROCEDURE

Scale Accuracy

Before weighing any animal, make sure that the hand of the scale is on zero when the scale is empty and that the scale is properly calibrated. To check the calibration, place a 20 g weight on the pan. If the hand does not read 20 g, turn the knob behind and underneath the pan until it points to 20 g.

To check the accuracy of the other types of scales, use a weight similar to the weight of the animal used in the experiment and adjust the needle if necessary. This should be done at the beginning of all experiments that require a scale.

Dose

The dose is 20 cc/kg 35% ethanol (ETOH), which means a mouse weighing 1,000 g will receive 20 cc 35% ETOH. Obviously a mouse does not weigh 1,000 g; it usually will weigh 15-40 g. If a mouse weighs, for example, 20 g, use the following formula to calculate the correct volume to give:

$$\frac{\text{volume of dose ordered}}{\text{1 kg body weight of mouse}} = \frac{\text{unknown volume to give (x)}}{\text{known weight of mouse}}$$

$$\frac{\text{20 cc ETOH}}{\text{1 kg}} = \frac{\text{x}}{\text{20 g}}$$

All measurements must be in the same units. Thus 1 kg must be changed to 1,000 g (or 20 g to 0.002 kg):

$$\frac{\text{20 cc ETOH}}{\text{1,000 g}} = \frac{\text{x}}{\text{20 g}}$$

$$(1{,}000\ \text{g})(\text{x}) = (20\ \text{cc})(20\ \text{g})$$

$$(1{,}000\ \text{g})(\text{x}) = (400\ \text{cc})(\text{g})$$

$$\text{x} = \frac{400\ \text{cc(g)}}{1{,}000\ \text{g}} = 0.4\ \text{cc}$$

Thus a 20 g mouse would receive 0.4 cc 35% ETOH. To calculate how much a 24 g mouse would receive:

$$\frac{\text{volume of dose ordered}}{\text{1 kg body weight}} = \frac{\text{unknown volume for 24 g mouse (x)}}{\text{body weight mouse (24 g)}}$$

$$\frac{\text{20 cc 35\% ETOH}}{\text{1 kg}} = \frac{\text{x}}{\text{24 g}}$$

$$\frac{\text{20 cc ETOH}}{\text{1,000 g}} = \frac{\text{x}}{\text{24 g}}$$

Cross multiplying:

$$(1{,}000\ \text{g})(\text{x}) = (20\ \text{cc})(24\ \text{g})$$

$$(1{,}000\ \text{g})(\text{x}) = (480\ \text{cc})(\text{g})$$

$$\text{x} = \frac{480\ \text{cc (g)}}{1{,}000\ \text{g}} = 0.48\ \text{cc}$$

This equation is used for a mouse of any weight, substituting its weight for the 24 g weight.

Filling Syringe

There are three main parts of the syringe: head, plunger, and barrel (Fig. 1.1). There are three parts of the feeding needle: hub, shaft, and blunted end (Fig. 1.2).

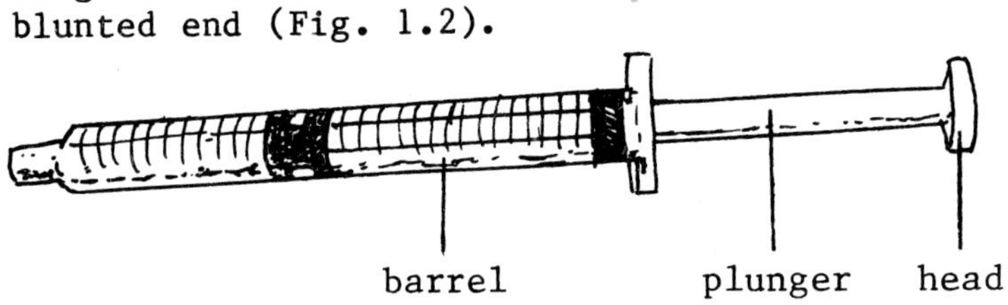

FIG. 1.1. Three main parts of syringe.

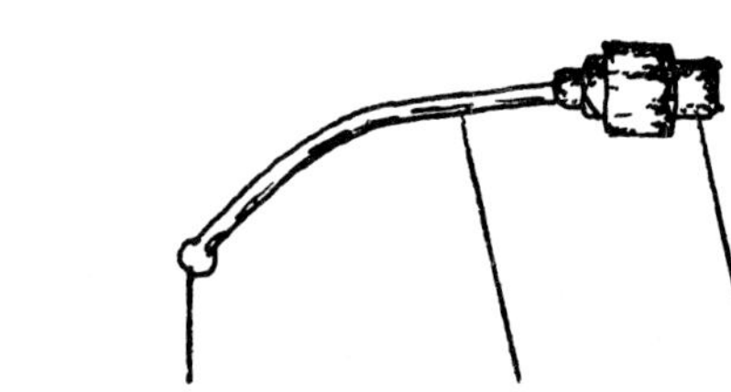

FIG. 1.2. Three parts of feeding needle.

Fasten the hub of the needle onto the hub of the syringe fairly tightly. To fill the syringe to 1 cc, use the beaker containing the tap water. Place the blunt end of the needle in the tap water and pull back the head of the plunger. The water will be drawn up into the syringe. Fill the syringe with water beyond 1 cc, to approximately 1.1 cc. To expel any trapped air in the syringe, invert the syringe so that the blunt end of the needle is pointing upward. Draw the plunger head gently downward another 0.1 cc, allowing more air to enter the needle and syringe. Tap the syringe gently on its barrel to encourage any trapped air bubbles to migrate to the top and gently depress the plunger so that liquid comes out of needle. Once liquid comes out, place the needle over the beaker to catch the excess expressed liquid and move the plunger end to rest on the 1.0 cc mark.

Practice filling the syringe several times.

Weighing and Handling Mouse

A mouse may be picked up by the base or middle of the tail for transfer to another cage for weighing. Never pick up a mouse by the end of the tail because all the mouse's weight will be suspended from the end of the tail. If the mouse struggles it is possible for the skin on the end of the tail to be ripped off.

Place the mouse on the scale. Record the weight on observation sheet 1.

On your student report (number 1), calculate the volume the mouse will receive and record the volume and the time on observation sheet 1. The water-treated control animals will be dosed first. Fill your syringe to the correct volume with tap water and set it aside.

Oral Administration

Grasp the mouse midtail with the left thumb and index finger (Fig. 1.3).

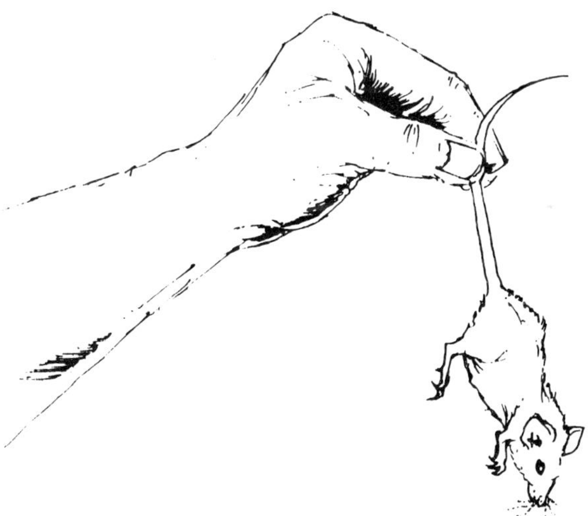

FIG. 1.3. Picking up mouse midtail.

FIG. 1.4. Mouse straining to escape.

(All instructions in this manual are for right-handed technicians; left-handed individuals should use the opposite hand.) Place the mouse on a grid or wire cage top where it can attempt to pull away from you, trying to escape (Fig. 1.4). Squeeze your other three left fingers around the base of the tail into your palm. Release your thumb and index finger but keep hold of the tail with

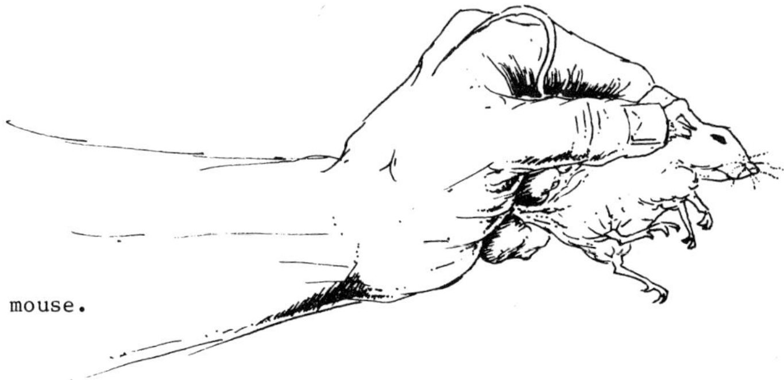

FIG. 1.5. Restrained mouse.

the other three fingers. Move your thumb and index finger to either side of the base of its skull and grasp the loose fur. Lift the animal up (Fig. 1.5).

Holding the mouse in your left hand, pick up your filled syringe in your right hand. Hold it as you would hold a dart, with the point of the needle corresponding to the position of the point of the dart.

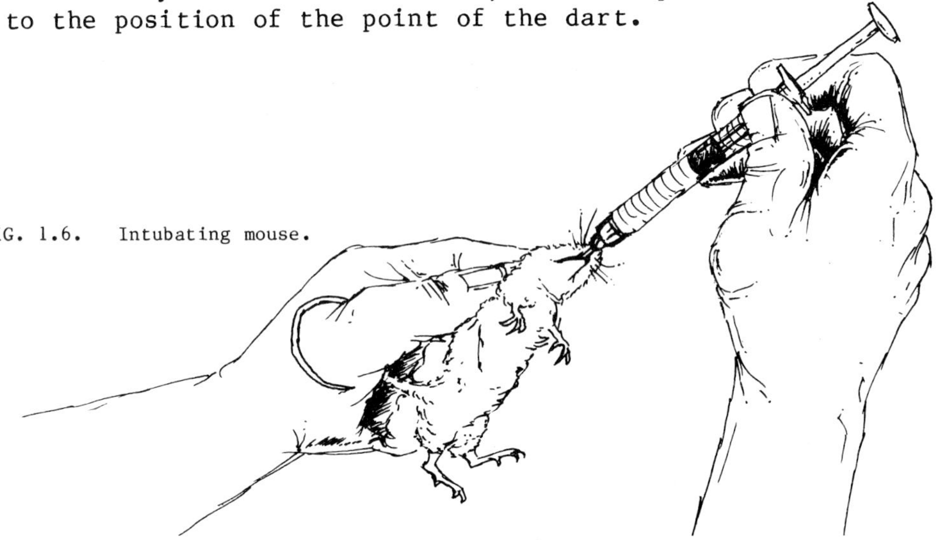

FIG. 1.6. Intubating mouse.

Place the tip of the feeding needle in the back of the mouse's throat (Fig. 1.6). Push the needle gently toward the stomach; it should slide down the esophagus into the stomach. Replace your index finger on the barrel of the syringe with the middle finger. With your index finger, depress the plunger head slowly, emptying the syringe contents into the stomach. If any obstruction is encountered, or if more than one-third of the needle is still showing, withdraw the needle and try again because you are not in the stomach. After emptying the syringe, withdraw the needle and place the mouse in the empty cage.

Dose a total of 10 mice with tap water, placing them all in one cage.

To dose the ETOH-treated animals, weigh the mouse and calculate the correct volume of ETOH to give (e.g., a 20 g mouse would receive 0.4 cc ETOH, a 24 g mouse 0.48 cc ETOH). Dose 10 mice with ETOH and place them all in a separate cage.

Results

Observe the 10 water-treated control mice periodically until all animals react normally (usually within a minute or so). Record your observations on observation sheet 1.

Observe the 10 ETOH-treated mice periodically (see Appendix D), comparing them with the water-treated mice, and record your observations on observation sheet 2. Be sure to note the times that you made your observations (e.g., 10 min, 20 min, etc.). Record any deaths during the 7 day period after dosing.

Complete your student report sheet.

EXP. 1: STUDENT REPORT

Name ______________________________

Date ______________________________

1. Calculate the volume of water the mouse should receive according to its weight (show your work).

2. What is 35% ethanol? Is it a central nervous system stimulant or depressant? Look it up in _The Pharmacological Basis of Therapeutics_, by Goodman and Gilman (see the Bibliography).

3. Why should the mice be without food 4 hr prior to dosing?

4. Why is it necessary to dose the same number of control mice?

5. Did any animal die? If so, why?

6. Define the following briefly:

   a. Ataxia--

   b. Coma--

   c. Corneal reflex--

   d. Cyanosis--

   e. Hypnosis--

   f. Lacrimation--

   g. Palpebral opening--

   h. Piloerection--

   i. Pain reflex--

   j. Righting reflex--

   k. Straub response--

   l. Vasoconstriction--

   m. Vasodilation--

EXP. 1: OBSERVATION SHEET 1
(water-treated mice)

Name______________________________

Date______________________________

Fill in the information below:

species__________ route__________ % sol./susp.__________
dose__________ drug__________ vehicle__________

Weigh the first animal and record the weight, calculate and administer the appropriate dose, and record the volume administered and the time. Repeat this procedure for all animals in the experiment.

Animal	1	2	3	4	5	6	7	8	9	10
Weight of animal (g)										
Volume administered (cc)										
Time dosed										

Observe the group as a whole (see Appendix D for definitions and observation techniques). Record the degree of response (0 = normal, + = slight, ++ = moderate, +++ = marked) after each item for the group. Take periodic observations until the group has returned to normal.

Time group observed					
Number dead/number dosed					
1. Activity increase					
2. Activity decrease					
3. Ataxia					
4. Coma					
5. Convulsion(s)					
6. Cyanosis					
7. Defecation, excessive					
8. Heart rate increase					
9. Heart rate decrease					
10. Irritability increase					
11. Irritability decrease					
12. Lacrimation increase					
13. Lacrimation decrease					
14. Muscle tone increase					
15. Muscle tone decrease					
16. Pain reflex loss					
17. Palpebral opening increase					
18. Palpebral opening decrease					
19. Piloerection					
20. Respiratory rate increase					
21. Respiratory rate decrease					
22. Righting reflex loss					
23. Salivation increase					
24. Startle response increase					
25. Startle response decrease					
26. Straub response					
27. Touch response increase					
28. Touch response decrease					
29. Tremors, twitches, jerks					
30. Urination increase					
31. Vasoconstriction increase					
32. Vasodilation increase					
33. Vocalization increase					

EXP. 1: OBSERVATION SHEET 2
(drug-treated mice)

Name____________________

Date____________________

Fill in the information below:

species__________ route__________ % sol./susp.__________
dose__________ drug__________ vehicle__________

Weigh the first animal and record the weight, calculate and administer the appropriate dose, and record the volume administered and the time. Repeat this procedure for all animals in the experiment.

Animal	1	2	3	4	5	6	7	8	9	10
Weight of animal (g)										
Volume administered (cc)										
Time dosed										

Observe the group as a whole (see Appendix D for definitions and observation techniques). Record the degree of response (0 = normal, + = slight, ++ = moderate, +++ = marked) after each item for the group. Take periodic observations until the group has returned to normal.

Time group observed					
Number dead/number dosed					
1. Activity increase					
2. Activity decrease					
3. Ataxia					
4. Coma					
5. Convulsion(s)					
6. Cyanosis					
7. Defecation, excessive					
8. Heart rate increase					
9. Heart rate decrease					
10. Irritability increase					
11. Irritability decrease					
12. Lacrimation increase					
13. Lacrimation decrease					
14. Muscle tone increase					
15. Muscle tone decrease					
16. Pain reflex loss					
17. Palpebral opening increase					
18. Palpebral opening decrease					
19. Piloerection					
20. Respiratory rate increase					
21. Respiratory rate decrease					
22. Righting reflex loss					
23. Salivation increase					
24. Startle response increase					
25. Startle response decrease					
26. Straub response					
27. Touch response increase					
28. Touch response decrease					
29. Tremors, twitches, jerks					
30. Urination increase					
31. Vasoconstriction increase					
32. Vasodilation increase					
33. Vocalization increase					

EXPERIMENT 2

Intramuscular Administration of Caffeine in the Mouse

INTRODUCTION

This experiment acquaints the student with removing liquids from glass-stoppered ampules, intramuscular (IM) injections in mice, and the effects of caffeine on the central nervous system.

In this experiment, 10 mice are injected intramuscularly with sterile water and 10 mice are injected with caffeine. The two groups are then compared to determine the effects of caffeine on mice.

MATERIALS AND EQUIPMENT (per student)

- caffeine, 250 mg/cc (1 ampule per 4 students)
- cylinder, graduated, 10 cc
- small rectangular or triangular metal file (1 per 4-8 students)
- sterile beaker with 500 cc sterile distilled water (per 10 students)
- sterile beaker, 25 cc
- needle, 20 gauge, 1/2 in.
- needle, 25 gauge, 1/2 in.
- syringe, 1 cc
- 20 mice, each approximately 20 g
- mouse scale and balance weights (per 2 students)
- 1 empty cage

PROCEDURE

Dose

The dose is 150 mg/kg and the volume 10 cc/kg, which means that every 1,000 g (1 kg) mouse should receive 150 mg in 10 cc sterile water. Since the vial of caffeine contains 250 mg/cc, the solution must be diluted to obtain a dose concentration of 150 mg/10 cc. To calculate the dilution use the following formula:

$$\frac{\text{known concentration}}{\text{known volume}} = \frac{\text{desired concentration}}{\text{unknown volume for dilution (x)}}$$

$$\frac{250 \text{ mg caffeine}}{1 \text{ cc}} = \frac{150 \text{ mg caffeine}}{x}$$

Cross multiplying:

$$(250 \text{ mg})(x) = (150 \text{ mg})(1 \text{ cc})$$

$$x = \frac{(150 \text{ mg})(\text{cc})}{250 \text{ mg}}$$

$$x = 0.6 \text{ cc}$$

After the dilution has been calculated, the ampule is opened.

Ampule

An ampule is a sealed glass container that usually holds only one dose of a drug that is given by injection. In this experiment, four students will share one ampule. For access to the solution, remove the top of the ampule. If there is liquid in the top of the ampule, tap the container at a 45° angle gently against a book. The liquid should drain out of the top section.

With a metal file, score (scratch) the glass ampule on one side along the blue line. Then apply pressure to the top and it will break off, giving access to the solution.

Each student will then make an individual dilution after using the 20 gauge needle and 1 cc syringe. Take 0.6 cc of the caffeine solution from the vial and place in the 10 cc graduated cylinder (express all trapped air bubbles in the syringe). Add enough sterile water (9.4 cc) to make a 10 cc volume, giving the desired concentration of 150 mg/10 cc. (The terminology is "to q.s. the solution to 10 cc.")

Then pour the diluted solution into the beaker and mark it "caffeine."

Volume

Each animal will receive 10 cc/kg of the diluted caffeine solution. This means each mouse will receive 10 cc of the diluted caffeine solution if it weighs 1,000 g. But most mice weigh 20 g, so use this formula:

$$\frac{\text{volume of dose ordered}}{\text{1 kg body weight of mouse}} = \frac{\text{unknown volume to give (x)}}{\text{known weight of mouse}}$$

$$\frac{10 \text{ cc}}{1{,}000 \text{ g}} = \frac{x}{20 \text{ g}}$$

$$(10 \text{ cc})(20 \text{ g}) = (1{,}000 \text{ g})(x)$$

$$\frac{(200 \text{ cc})(\text{g})}{1{,}000 \text{ g}} = x$$

$$0.2 \text{ cc} = x$$

Thus a 20 g mouse will receive 0.2 cc of the diluted caffeine solution. The same calculation is repeated for each mouse, substituting its weight in the equation.

On the student report (number 5), calculate the volume to give a 24 g mouse (answer: 0.24 cc).

Weighing

Check your scale for accuracy and adjust if necessary (see Exp. 1). Weigh the first mouse and record the weight on the observation sheet. Calculate the volume to give according to this weight and record it on the observation sheet.

IM Injection

The 25 gauge beveled needle is similar to the feeding needle used in Exp. 1 except that it has a sharp end, called a bevel (Fig. 2.1).

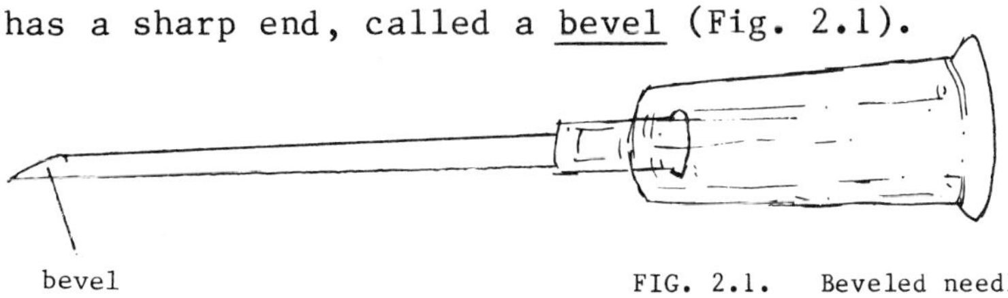

FIG. 2.1. Beveled needle for injections.

The control animals will be dosed first with sterile distilled water. Put the needle on the syringe and fill the syringe to the correct volume (a 20 g mouse will receive 0.2 cc).

Students work in pairs to dose the animals. Have your lab partner restrain the mouse by the method described in Exp. 1. Hold one of its back feet in your left hand and pick up your syringe in your right hand (Fig. 2.2).

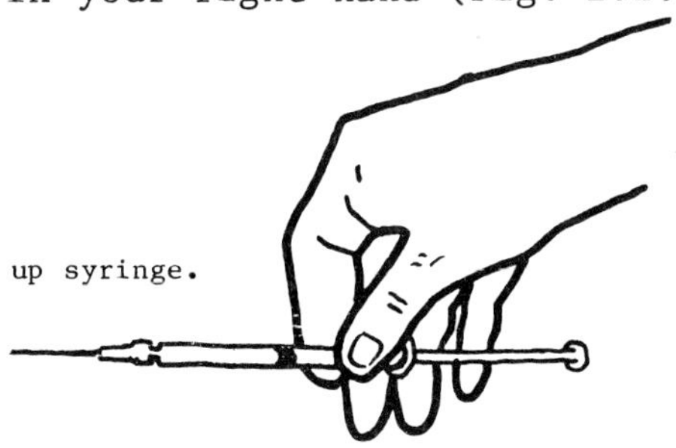

FIG. 2.2. Picking up syringe.

Insert the needle at a 90° angle into the semitendinosus muscle about 1/4 in. (Fig. 2.3). Normally when you give an IM injection, slight negative pressure should be exerted on the plunger with your right hand to make sure the needle is not in a blood vessel (blood will flow into the syringe if the needle is in a blood vessel). With IM injections in rats and mice this is not necessary, however, because their blood vessels are too small.

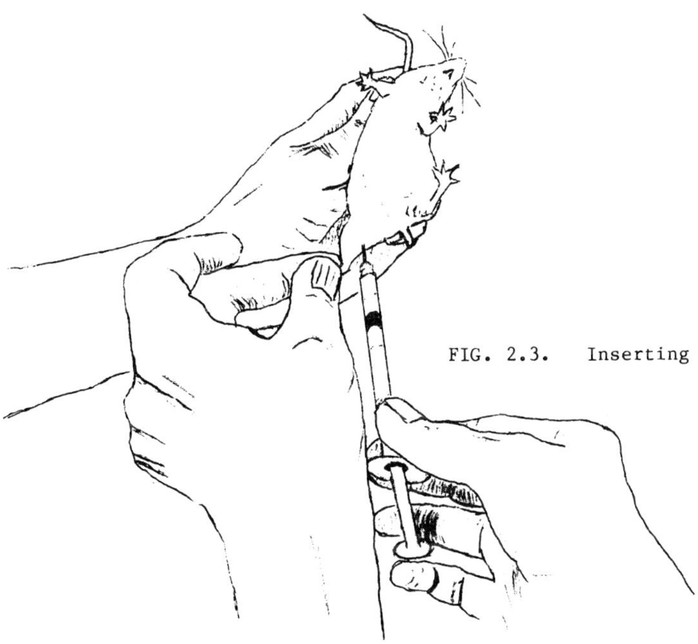

FIG. 2.3. Inserting needle into semitendinosus muscle.

Inject the dose, withdraw the needle, and release the mouse in an empty cage. Repeat the procedure with 9 more mice. Then partners reverse roles and repeat the procedure with the partner's group of 10 mice.

Dose the remaining two groups of 10 mice with the diluted caffeine solution. Calculate the volume to give each mouse according to its weight as you did with the control mice.

Results

Students observe their individual groups of mice.

Observe the 10 caffeine-treated mice periodically, comparing them with the water-treated mice, and record your observations on the observation sheet.

Complete your student report sheet.

EXP. 2: STUDENT REPORT

Name______________________________

Date______________________________

1. What is caffeine?

2. What main organs does caffeine affect?

3. Why was food not withheld from these animals before dosing?

4. What does _q_._s_. mean?

5. Calculate the volume to give a 24 g mouse (show your work).

6. Calculate the dilution needed for a 600 mg/kg caffeine dose and a 10 cc/kg volume (show your work).

EXP. 2: OBSERVATION SHEET

Name______________________________

Date______________________________

Fill in the information below:

species__________ route__________ % sol./susp.__________
dose_____________ drug___________ vehicle______________

Weigh the first animal and record the weight, calculate and administer the appropriate dose, and record the volume administered and the time. Repeat this procedure for all animals in the experiment.

Animal	1	2	3	4	5	6	7	8	9	10
Weight of animal (g)										
Volume administered (cc)										
Time dosed										

Observe the group as a whole (see Appendix D for definitions and observation techniques). Record the degree of response (0 = normal, + = slight, ++ = moderate, +++ = marked) after each item for the group. Take periodic observations until the group has returned to normal.

Time group observed					
Number dead/number dosed					
1. Activity increase					
2. Activity decrease					
3. Ataxia					
4. Coma					
5. Convulsion(s)					
6. Cyanosis					
7. Defecation, excessive					
8. Heart rate increase					
9. Heart rate decrease					
10. Irritability increase					
11. Irritability decrease					
12. Lacrimation increase					
13. Lacrimation decrease					
14. Muscle tone increase					
15. Muscle tone decrease					
16. Pain reflex loss					
17. Palpebral opening increase					
18. Palpebral opening decrease					
19. Piloerection					
20. Respiratory rate increase					
21. Respiratory rate decrease					
22. Righting reflex loss					
23. Salivation increase					
24. Startle response increase					
25. Startle response decrease					
26. Straub response					
27. Touch response increase					
28. Touch response decrease					
29. Tremors, twitches, jerks					
30. Urination increase					
31. Vasoconstriction increase					
32. Vasodilation increase					
33. Vocalization increase					

EXPERIMENT 3

Intravenous Administration of Sodium Pentobarbital in the Mouse

INTRODUCTION

This experiment acquaints the student with filling a syringe from a stoppered vial, giving intravenous (IV) injections, and the behavioral and/or physiological effects of sodium pentobarbital (Nembutal) in mice.

In this experiment, 10 mice are injected intravenously with sodium pentobarbital and 10 mice are untreated. The two groups are then compared to determine the effects of sodium pentobarbital in mice.

MATERIALS AND EQUIPMENT (per student)

- syringe, 1 cc
- needle, 24 gauge, 5/8 in.
- 2 cotton balls
- 50 cc bottle of isopropyl alcohol, 70% (per 20 students)
- sodium pentobarbital (Nembutal), 60 mg/ml (100 cc vial per 20 students)
- sterile graduated cylinder, 50 cc
- sterile beaker with 500 cc sterile distilled water (per 10 students)
- mouse scale and balance weights (per 2 students)
- 20 mice, each approximately 20 g
- wire mesh cage
- mouse restrainer (if wire mesh cage not used)
- empty cage (if wire mesh cage not used)

PROCEDURE

Dose

The dose is 35 mg/kg. But different stocks and/or strains of mice can deviate somewhat from the normal reaction. It is possible that the dose might have to be adjusted by the instructor.

Using this dose, on the student report (number 2), calculate the correct milligrams of pentobarbital to give a 20 g mouse.

Volume

The concentration of pentobarbital in the vial is 60 mg/cc. On the student report (number 3), calculate the volume to give a 20 g mouse (answer: 0.012 cc).

Find 0.012 cc on your syringe. Since you cannot accurately give this small a dose, you must dilute the concentration and give a larger volume. To do this, withdraw 1 cc of the concentration from the vial.

Filling Syringe from Vial

Place the needle on the syringe. Clean the top of the stoppered vial with a cotton ball dipped in 70% alcohol for aseptic procedure.

Since the stoppered vial is airtight, it is easier to withdraw liquid if air is injected into the vial first. Withdraw the plunger to 1 cc to allow air to enter the syringe.

Hold the vial in your left hand, with the bottom of the vial facing upward between your thumb and first two fingers. The remaining two fingers are free (Fig. 3.1).

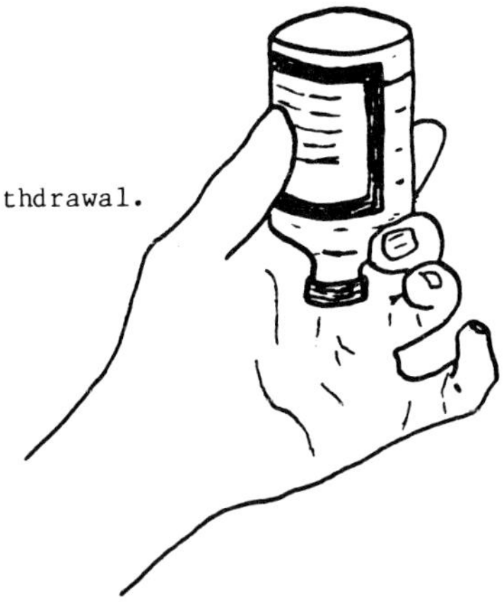

FIG. 3.1. Holding vial for liquid withdrawal.

Pick up the syringe in your right hand, supporting the plunger with your little finger as in Fig. 2.2.

Insert the syringe needle into the vial, making sure the beveled end is in the fluid. Clasp your free two left fingers around the syringe barrel for support (Fig. 3.2).

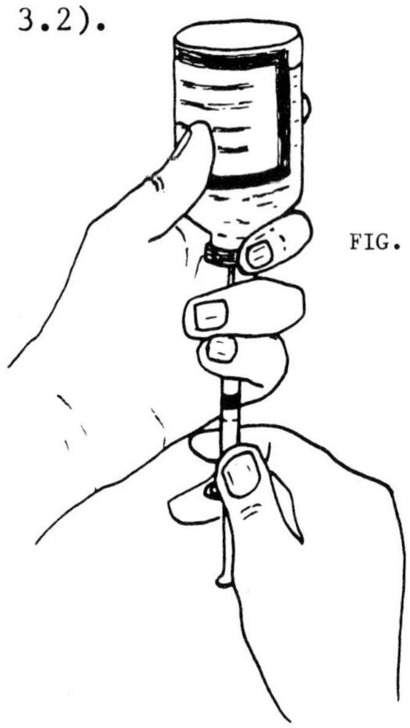

FIG. 3.2. Inserting syringe into vial.

Drop your right thumb to the plunger head and inject the air into the vial. Then, still holding the vial and syringe in your left hand, use your right thumb and forefinger to withdraw the plunger head to the 1 cc mark.

Then slide the left hand back to the barrel and support the plunger again with the little finger. Withdraw the needle and syringe from vial.

Dilution

Add the 1 cc of pentobarbital to the graduated cylinder. Carefully pour in the sterile distilled water to q.s. to 50 cc (you are diluting 60 mg/1 cc to 60 mg/50 cc).

Calculate how many milligrams of pentobarbital are now in 1 cc:

$$\frac{\text{x mg pentobarbital}}{\text{1 cc}} = \frac{\text{60 mg pentobarbital}}{\text{50 cc sterile water}}$$

Cross multiplying:

$$(50 \text{ cc})(x) = (60 \text{ mg})(1 \text{ cc})$$

$$x = 1.2 \text{ mg pentobarbital}$$

Thus 1 cc of the new dilution will have 1.2 mg pentobarbital.

Now consider your dose. You want to give each 20 g mouse 0.7 mg, and you how have a dilution of 1.2 mg/cc. On your student report (number 4), calculate the volume (answer: 0.58 cc).

Remember you must also correct the volume for any animal weighing more or less than 20 g (e.g., a 25 g mouse would receive 0.73 cc of the diluted pentobarbital solution).

Weighing

Check your scale for accuracy and adjust as necessary (see Exp. 1). Weigh your animal, calculate the volume needed according to its weight, fill your syringe, and lay it aside. Record the weight, volume, and time on the observation sheet.

IV Injection

Restraint for IV. Place the mouse in the wire cage or mouse restrainer but keep hold of the tail. Thread the tail through the mesh hole (or restrainer hole) so that the body is on the inside and the tail is on the outside. Hold the tail (about midtail) with your left hand.

Injection site. There are two arteries in the mouse's tail, dorsal and ventral (top and bottom), and two veins, right and left lateral (side). These veins are located just below the surface of the skin. With your thumb and index finger rotate the tail 45° (either clockwise or counterclockwise). After rotating the tail the blood vessel you see dorsally is the vein (Fig. 3.3).

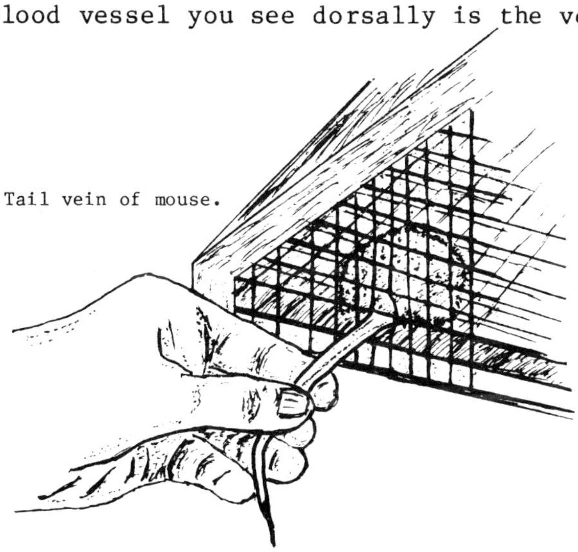

FIG. 3.3. Tail vein of mouse.

Injection technique. Holding the rotated tail parallel to the mouse's body, pick up the loaded syringe and hold it as you did for the IM injection (Fig. 2.2). Make sure the beveled edge of the needle is facing upward. Insert your needle at a 20° angle into the tail vein about 1/4 in. (Fig. 3.4). Slowly inject the pentobarbital into the vein.

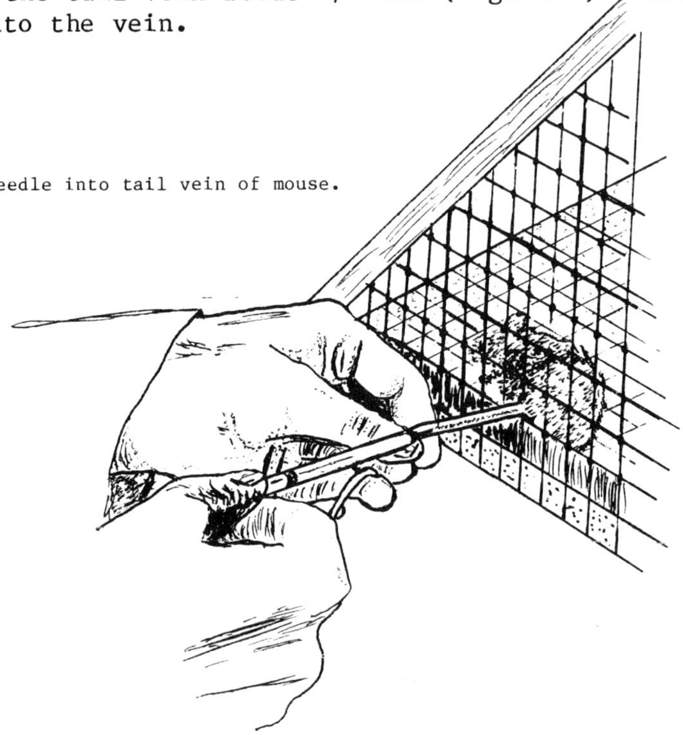

FIG. 3.4. Inserting needle into tail vein of mouse.

If the fluid doesn't flow easily you're not in the vein and should stop immediately and remove your needle. As you pull the needle out, place your thumb over the injection site and apply slight pressure for a minute or so to prevent hemorrhage. Try again further up the tail about 1/2 in. (cranially) or in the vein on the other side of the tail.

Repeat this procedure with 9 more mice. If your needle was completely in the vein, all the mice should show the same responses and should be anesthetized.

Observe and compare results of the 10 pentobarbital-treated animals with the 10 untreated control mice.

Record your observations of the drug-treated mice on the observation sheet. Be sure to note duration of anesthesia (how long the animals were anesthetized).

Complete your student report sheet.

EXP. 3: STUDENT REPORT

Name______________________________

Date______________________________

1. What type of drug is sodium pentobarbital? Look it up in _The Pharmacological Basis of Therapeutics_, by Goodman and Gilman (see the Bibliography).

2. Calculate the milligrams to give a 20 g mouse (show your work).

3. Calculate the volume to give a 20 g mouse (show your work).

4. Calculate the volume needed to give a 20 g mouse 0.7 mg using a dilution of 1.2 mg/cc (show your work).

5. Calculate the following doses and volumes (show your work):

   a. For a desired dose of 50 mg/kg pentobarbital, how many milligrams should a 20 g mouse receive?

   b. At a dose of 50 mg/kg and a pentobarbital concentration of 2 mg/cc in the vial, what volume should this 20 g mouse receive?

   c. Can you accurately give the mouse this volume by syringe?_____

   d. At the dose and volume in 5b, what volume should a 20 g mouse receive?

6. The generally recommended anesthesia dose of sodium pentobarbital IV in mice is 35 mg/kg. Why would some mice have an anesthesia dose of 40 mg/kg?

7. Give three other factors that might cause an anesthesia dose to be higher than normal.

EXP. 3: OBSERVATION SHEET

Name____________________________

Date____________________________

Fill in the information below:

species__________ route__________ % sol./susp.__________
dose__________ drug__________ vehicle__________

Weigh the first animal and record the weight, calculate and administer the appropriate dose, and record the volume administered and the time. Repeat this procedure for all animals in the experiment.

Animal	1	2	3	4	5	6	7	8	9	10
Weight of animal (g)										
Volume administered (cc)										
Time dosed										

Observe the group as a whole (see Appendix D for definitions and observation techniques). Record the degree of response (0 = normal, + = slight, ++ = moderate, +++ = marked) after each item for the group. Take periodic observations until the group has returned to normal.

Time group observed					
Number dead/number dosed					
1. Activity increase					
2. Activity decrease					
3. Ataxia					
4. Coma					
5. Convulsion(s)					
6. Cyanosis					
7. Defecation, excessive					
8. Heart rate increase					
9. Heart rate decrease					
10. Irritability increase					
11. Irritability decrease					
12. Lacrimation increase					
13. Lacrimation decrease					
14. Muscle tone increase					
15. Muscle tone decrease					
16. Pain reflex loss					
17. Palpebral opening increase					
18. Palpebral opening decrease					
19. Piloerection					
20. Respiratory rate increase					
21. Respiratory rate decrease					
22. Righting reflex loss					
23. Salivation increase					
24. Startle response increase					
25. Startle response decrease					
26. Straub response					
27. Touch response increase					
28. Touch response decrease					
29. Tremors, twitches, jerks					
30. Urination increase					
31. Vasoconstriction increase					
32. Vasodilation increase					
33. Vocalization increase					

EXPERIMENT 4

Intraperitoneal Administration of Tubocurarine in the Mouse

INTRODUCTION

This experiment acquaints the student with giving intraperitoneal (IP) injections in mice and with the behavioral and/or physiological effects of the neuromuscular blocking agent, tubocurarine, in mice.

In this experiment, 10 mice are injected intraperitoneally with tubocurarine and 10 mice are untreated. The two groups are then compared to determine the effects of tubocurarine on mice.

Tubocurarine causes striated muscle paralysis. Direct stimulation of normal muscle motor nerves usually produces muscle contraction. In tubocurarine-treated animals direct stimulation of muscle nerves causes no reaction, but if the muscle itself is stimulated there is contraction. It has been shown that tubocurarine competes with and blocks acetylcholine, a natural myoneural substance that helps to transport impulses from the nerve to the muscle at the myoneural junction.

Tubocurarine was originally extracted from the South American vine *Chondodendron tomentosum*. Crude extracts of this vine were used by South American Indians to paralyze wild game via blowguns. Tubocurarine is now used as an adjunct muscle relaxant during surgery, allowing the dose of general anesthesia to be reduced.

MATERIALS AND EQUIPMENT (per student)

- tubocurarine, 3 mg/cc (1 vial per 20 students)
- 2 syringes, 1 cc
- needle, 20 gauge, 1 in.
- sterile graduated cylinder, 100 cc
- sterile beaker with 100 cc sterile distilled water
- 20 mice, each approximately 20 g
- mouse scale and balance weights (per 2 students)
- needle, 25 gauge, 1/4 in.
- empty cage

PROCEDURE

Dose

The dose is 0.25 mg/kg. (The reported LD_{50}, the lethal dose causing 50% of the test animals to die, is 0.15 mg/kg in mice. However, the author found the reported LD_{50} dose did not produce any deaths or even symptoms when it was tested, probably because a different strain of mice was used. Since there appear to be significant strain differences, it is possible that the recommended dose of 0.25 mg/kg may have to be adjusted for a specific strain of mice. It is recommended that the instructor test the suggested dose on several mice and make the necessary adjustments before the students perform this experiment.)

On the student report (number 4), calculate the milligrams of tubocurarine to give a 20 g mouse (answer: 0.0005 mg).

Volume

The concentration in the vial is 3 mg/cc. On the student report (number 5), calculate the volume to give a 20 g mouse (answer: 0.000166 cc).

This dose is obviously too small a volume to administer. Thus, the concentration must be diluted. Take 0.17 cc of tubocurarine and q.s. to 100 cc with sterile distilled water (place 0.17 cc of tubocurarine in the graduated cylinder using the 1 cc syringe and 20 gauge needle; add enough sterile distilled water to make a 100 cc volume). Mix the dilution well.

Weighing

Check your scale for accuracy and adjust if necessary (see Exp. 1). Weigh the first mouse and record the weight on the observation sheet. If it weighs more or less than 20 g, calculate the volume to give according to its weight and record it on the observation sheet.

IP Injection

Using the second 1 cc syringe and the 25 gauge, 1/4 in. needle, fill the syringe to the correct volume and lay it aside.

Restrain the mouse as previously done for oral dosing (Figs. 1.3-1.5). Then pick up the syringe in your right hand and insert the needle about 1/8 in. into the peritoneal cavity, midway between the tail and the umbilicus (Fig. 4.1).

Repeat the procedure with the remaining 9 mice.

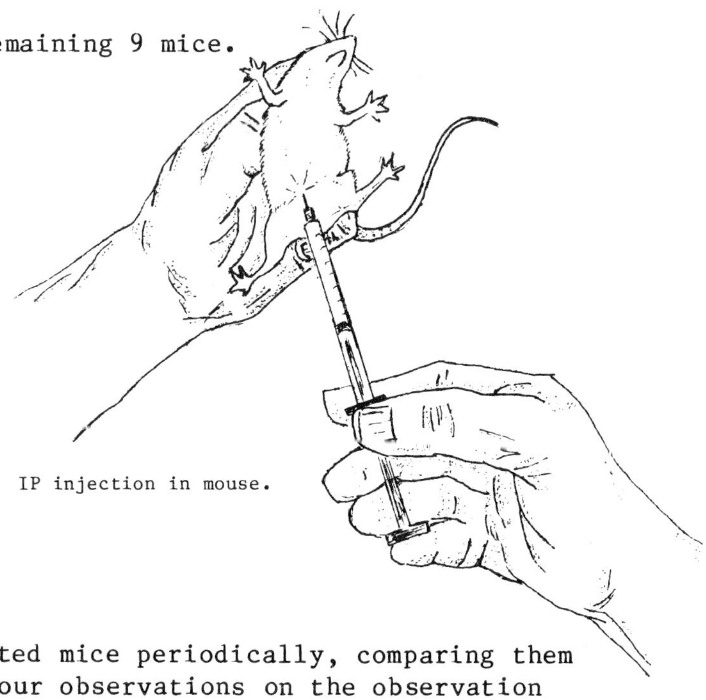

FIG. 4.1. IP injection in mouse.

Results

Observe the 10 tubocurarine-treated mice periodically, comparing them with the 10 untreated mice. Record your observations on the observation sheet. Be sure to note the onset and duration of the effects.

Complete your student report sheet.

EXP. 4: STUDENT REPORT

Name______________________________

Date______________________________

1. What type of drug is tubocurarine? Look it up in the Physicians' Desk Reference (see the Bibliography).

2. What organ(s) does it affect?

3. Can the tubocurarine-treated animal feel pain if its muscles are directly stimulated?

4. Calculate the milligrams of tubocurarine to give a 20 g mouse (show your work).

5. Calculate the volume to give a 20 g mouse (show your work).

6. Calculate the following doses and volumes (show your work):

   a. For a 0.3 mg/kg dose of tubocurarine, how many milligrams should a 20 g mouse receive?

   b. At the same dose but with a concentration of 2 mg/cc, what volume should this 20 g mouse receive?

   c. Can you accurately give a mouse this volume by syringe? ____________

   d. Using the dose and concentration in 4b, calculate the dilution and corrected volume to give a 20 g mouse.

EXP. 4: OBSERVATION SHEET

Name________________________

Date________________________

Fill in the information below:

species__________ route__________ % sol./susp.__________
dose__________ drug__________ vehicle__________

Weigh the first animal and record the weight, calculate and administer the appropriate dose, and record the volume administered and the time. Repeat this procedure for all animals in the experiment.

Animal	1	2	3	4	5	6	7	8	9	10
Weight of animal (g)										
Volume administered (cc)										
Time dosed										

Observe the group as a whole (see Appendix D for definitions and observation techniques). Record the degree of response (0 = normal, + = slight, ++ = moderate, +++ = marked) after each item for the group. Take periodic observations until the group has returned to normal.

Time group observed					
Number dead/number dosed					
1. Activity increase					
2. Activity decrease					
3. Ataxia					
4. Coma					
5. Convulsion(s)					
6. Cyanosis					
7. Defecation, excessive					
8. Heart rate increase					
9. Heart rate decrease					
10. Irritability increase					
11. Irritability decrease					
12. Lacrimation increase					
13. Lacrimation decrease					
14. Muscle tone increase					
15. Muscle tone decrease					
16. Pain reflex loss					
17. Palpebral opening increase					
18. Palpebral opening decrease					
19. Piloerection					
20. Respiratory rate increase					
21. Respiratory rate decrease					
22. Righting reflex loss					
23. Salivation increase					
24. Startle response increase					
25. Startle response decrease					
26. Straub response					
27. Touch response increase					
28. Touch response decrease					
29. Tremors, twitches, jerks					
30. Urination increase					
31. Vasoconstriction increase					
32. Vasodilation increase					
33. Vocalization increase					

EXPERIMENT 5

Intravenous Administration of Thiamylal in the Rabbit

INTRODUCTION

This experiment acquaints the student with restraining and giving intravenous (IV) injections in rabbits.

In this experiment, one rabbit is injected IV with thiamylal and a second rabbit is injected with sterile water. The two animals are then compared to determine the effects of thiamylal in rabbits.

MATERIALS AND EQUIPMENT (per student)

- sodium thiamylal (Bio-tal, Surital)(1 g vial per 10 students)
- syringe, 20 cc (per 10 students)
- needle, 20 gauge, 1 in. (per 10 students)
- sterile beaker with 100 cc sterile distilled water
- rabbit scale and balance weights (1 per 5 students)
- 2 rabbits
- 2 stockades
- syringe, 3 cc
- needle, 25 gauge, 5/8 in.

PROCEDURE

Dose

The dose is 15 mg/kg, to effect. (To effect means to give slowly until the desired effect is achieved--in this case, anesthesia. The complete dose may not be necessary.)

On the student report (number 2), calculate the milligrams of thiamylal to give a 4 kg rabbit (answer: 60 mg).

Concentration

Since thiamylal is available as 1 g powder in a large vial, the concentration made can vary from a 0.5% to a 4% solution, depending on the volume of sterile water added. The concentration desired in this case is a 2.5% solution. Therefore add 40 cc sterile water to the vial. (Since 10 students are sharing one vial, one student makes the dilution.) Use the 20 cc syringe and 20 gauge needle to add the sterile water. Mix the solution well by agitating the vial.

Volume

On the student report (number 3) calculate the volume to give a 4 kg rabbit (answer: 2.4 cc to effect).

Weighing

Check your scale for accuracy and adjust as necessary (see Exp. 1). Weigh the one rabbit and record the weight on observation sheet 1. Calculate the volume to give according to the weight and record it on the observation sheet.

FIG. 5.1. Rabbit stockade for restraint.

IV Injection

Water-treated (control) rabbit. Using the 3 cc syringe and 25 gauge needle (see Exp. 4), fill your syringe with sterile water and lay it aside.

Open the front of the stockade (Fig. 5.1) and place the rabbit inside. With your right hand take its ears, quickly but gently pull its head beyond the opening, and quickly close the stockade door. If the animal jumps around, place a hand on its back to quiet it. Rabbits can easily injure their spines.

Pluck the hair from the right ear for about 2 in. along the lateral edge from the middle of the ear toward the head (cranially). The marginal ear vein should be visible along the edge of the ear.

With your left hand hold the ear approximately halfway between the head and the tip of the ear. The marginal ear vein should be between your index finger and thumb. Keep the ear parallel to the floor. Pick up the syringe with the beveled end of the needle facing upward and insert the needle at a 20° angle through the skin and into the vein about 1/4 in. (Fig. 5.2).

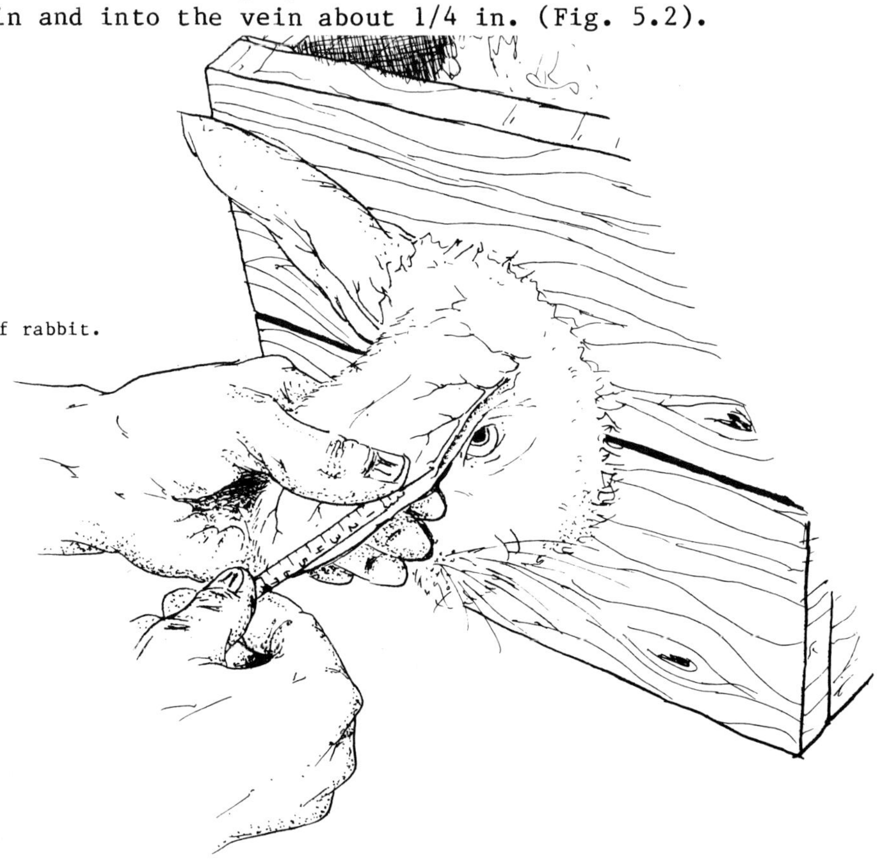

FIG. 5.2. IV injection in ear of rabbit.

Slowly inject the fluid into the ear. You should be able to see the fluid flow into and up the vein, pushing the blood ahead of it. If the fluid doesn't flow easily, you are not in the vein and you should stop immediately and remove the needle. As you pull the needle out, place your thumb over the injection site and apply light pressure to prevent hemorrhage. Try again about 1/2 in. further up the ear (cranially) or in the other ear.

After giving the water injection, the rabbit may be left in the stockade until returned to the cage.

Thiamylal-treated rabbit. Weigh the second rabbit and calculate the exact volume to give. Fill your syringe with thiamylal and lay it aside. Place the animal in the stockade. You must observe the animal very carefully as you give the injection as too much anesthesia will kill it.

Slowly give about one-half the dose, periodically checking the rabbit's wink and pain reflexes. (The wink reflex is checked by touching the median corner of its eye with your finger; an anesthetized animal will not wink. The pain reflex is checked by pinching a pad of its foot; an anesthetized animal will show no reaction.)

Continue giving the dose slowly to effect. Remove the needle and release the stockade latch quickly so the animal will not strangle.

Results

Record your observations of the thiamylal-treated rabbit on observation sheet 2. Be sure to note the onset and duration of anesthesia.

Complete your student report sheet.

EXP. 5: STUDENT REPORT

Name________________________________

Date________________________________

1. What type of drug is sodium thiamylal? Look it up in the Physicians' Desk Reference (see the Bibliography). Is it considered an ultrashort, short, intermediate, or long-acting anesthesia?

2. Calculate the milligrams of thiamylal to give a 4 kg rabbit (show your work).

3. Calculate the volume to give a 4 kg rabbit (show your work).

4. For a 25 mg/kg dose, using a 2% concentration, calculate the following (show your work):

   a. What volume of sterile distilled water should you add to the thiamylal vial?

b. What volume of the mixture in 4a should you give a 3 kg rabbit?

EXP. 5: OBSERVATION SHEET 1
(water-treated rabbit)

Name____________________

Date____________________

Fill in the information below:

species__________ route__________ % sol./susp.__________
dose__________ drug__________ vehicle__________

Weigh the animal and record the weight, calculate and administer the appropriate dose, and record the volume administered and the time.

Weight of animal (kg) __________
Volume administered (cc) __________
Time dosed __________

Observe the animal (see Appendix D for definitions and observation techniques). Record the degree of response (0 = normal, + = slight, ++ = moderate, +++ = marked) after each item. Take periodic observations until the animal has returned to normal.

Time observed					
Number dead/number dosed					
1. Activity increase					
2. Activity decrease					
3. Ataxia					
4. Coma					
5. Convulsion(s)					
6. Corneal reflex loss					
7. Cyanosis					
8. Defecation, excessive					
9. Heart rate increase					
10. Heart rate decrease					
11. Irritability increase					
12. Irritability decrease					
13. Lacrimation increase					
14. Lacrimation decrease					
15. Muscle tone increase					
16. Muscle tone decrease					
17. Pain reflex loss					
18. Palpebral opening increase					
19. Palpebral opening decrease					
20. Piloerection					
21. Respiratory rate increase					
22. Respiratory rate decrease					
23. Righting reflex loss					
24. Salivation increase					
25. Startle response increase					
26. Startle response decrease					
27. Straub response					
28. Touch response increase					
29. Touch response decrease					
30. Tremors, twitches, jerks					
31. Urination increase					
32. Vasoconstriction increase					
33. Vasodilation increase					
34. Vocalization increase					

EXP. 5: OBSERVATION SHEET 2
(drug-treated rabbit)

Name______________________________

Date______________________________

Fill in the information below:

species__________ route__________ % sol./susp.__________
dose__________ drug__________ vehicle__________

Weigh the animal and record the weight, calculate and administer the appropriate dose, and record the volume administered and the time.

Weight of animal (kg) ____________
Volume administered (cc) ____________
Time dosed ____________

Observe the animal (see Appendix D for definitions and observation techniques). Record the degree of response (0 = normal, + = slight, ++ = moderate, +++ = marked) after each item. Take periodic observations until the animal has returned to normal.

Time observed					
Number dead/number dosed					
1. Activity increase					
2. Activity decrease					
3. Ataxia					
4. Coma					
5. Convulsion(s)					
6. Corneal reflex loss					
7. Cyanosis					
8. Defecation, excessive					
9. Heart rate increase					
10. Heart rate decrease					
11. Irritability increase					
12. Irritability decrease					
13. Lacrimation increase					
14. Lacrimation decrease					
15. Muscle tone increase					
16. Muscle tone decrease					
17. Pain reflex loss					
18. Palpebral opening increase					
19. Palpebral opening decrease					
20. Piloerection					
21. Respiratory rate increase					
22. Respiratory rate decrease					
23. Righting reflex loss					
24. Salivation increase					
25. Startle response increase					
26. Startle response decrease					
27. Straub response					
28. Touch response increase					
29. Touch response decrease					
30. Tremors, twitches, jerks					
31. Urination increase					
32. Vasoconstriction increase					
33. Vasodilation increase					
34. Vocalization increase					

Oral Administration of Dextroamphetamine Sulfate in the Rat

INTRODUCTION

This experiment acquaints the student with handling, restraining, and oral administration (PO) in rats.

In this experiment, 10 rats are intubated (dosed orally) with dextroamphetamine sulfate and 10 rats are intubated with water. The two groups are then compared to determine the effects of dextroamphetamine sulfate in rats.

MATERIALS AND EQUIPMENT (per student)

dextroamphetamine sulfate (Dexedrine), 5 mg tablets, 2 tablets
mortar and pestle, 3 in.
funnel, glass, 3 in.
spatula, micro
graduated cylinder, 20 cc
beaker, 25 cc
beaker with 50 cc tap water
rat scale and balance weights (per 2 students)
20 rats, each approximately 60 g (held without food 4 hr before experiment)
feeding needle, 16 gauge, 3 in.
syringe, 1 cc
2 empty cages

PROCEDURE

Dose

The dose is 10 mg/kg. On the student report (number 3), calculate the milligrams of Dexedrine to give to a 100 g rat (answer: 0.10 mg). (Review Exps. 1 and 2 if you are unsure of how to do the calculations.)

The concentration of Dexedrine needed is 10 mg/20 cc water. Place two tablets in the mortar and crush them with the pestle.

Place the funnel into the graduated cylinder. Scrape all the powder into the funnel with the spatula and add enough tap water to <u>q.s.</u> the solution to 15 cc in the graduated cylinder. Transfer the solution into the 25 cc beaker. Rinse the cylinder out with an additional 5 cc tap water and transfer it into the 25 cc beaker.

Volume

Each animal will receive 10 cc/kg. On the student report (number 4), calculate the volume to give a 100 g rat (answer: 1 cc) (Review Exps. 1 and 2, if necessary.) The control animals will receive the same volume of tap water.

Weighing

Check your scale for accuracy and adjust as necessary (see Exp. 1). Pick up rats gently by midtail. Weigh the first rat and record the weight on the first observation sheet. Calculate the volume to give according to this weight and record it on the observation sheet.

Restraint and Oral Administration

Dose the control animals first. Place the feeding needle on the syringe, fill it to the correct volume with tap water, and lay it aside. Record the time on observation sheet 1.

There are several methods of restraining rats. The method usually preferred is as follows. Pick up the rat gently by midtail with your left hand. Gently place it in your right hand, with its head in front of your index finger and thumb (Fig. 6.1). Hold it loosely.

FIG. 6.1. Positioning rat gently in right hand.

Put the index finger and thumb of your left hand just behind and below its ears and grasp the skin firmly (Fig. 6.2). Gently turn the rat upside down (stomach up) and clasp the loose skin along its back with your remaining three fingers (Fig. 6.3).

FIG. 6.2. Grasping skin behind rat's neck while holding its lower body.

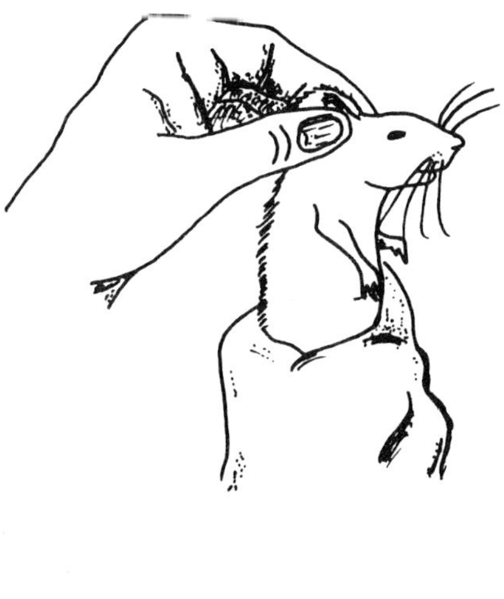

FIG. 6.3. Securing hold with remaining three fingers on back of rat.

With your right hand, pick up the loaded syringe and gently slide the feeding needle through the rat's mouth, down the esophagus, and into the stomach. Empty the syringe contents slowly into the stomach and withdraw the needle (Fig. 6.4).

FIG. 6.4. Intubating rat; syringe is held like a dart.

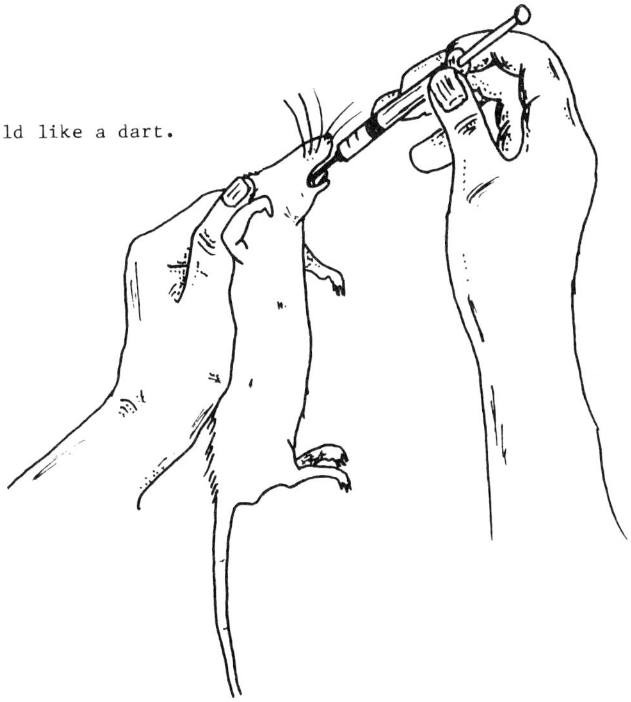

Place the dosed control rat in the empty cage. Repeat this procedure with 9 additional rats.

Dose the remaining 10 rats with Dexedrine. Stir the solution of Dexedrine and fill the feeding needle with the correct volume of Dexedrine solution. Proceed as you did with the control animals, putting the 10 Dexedrine-dosed animals into the second cage.

Results

Observe the treated animals periodically and compare them with the control animals until all animals react normally.

Record your observations of both groups on the observation sheets.

Complete the student report sheet.

EXP. 6: STUDENT REPORT

Name____________________________

Date____________________________

1. What type of drug is Dexedrine? Look it up in _The Pharmacological Basis of Therapeutics_, by Goodman and Gilman (see the Bibliography).

2. How does it work?

3. Calculate the milligrams of Dexedrine to give a 100 g rat (show your work).

4. Calculate the volume to give a 100 g rat (show your work).

5. For a 35 mg/kg dose of Dexedrine, calculate how many milligrams a 100 g rat should receive (show your work).

6. At a 35 mg/kg dose and a volume of 10 cc/kg, calculate how many 5 mg Dexedrine tablets and what volume of tap water would be needed (show your work).

EXP. 6: OBSERVATION SHEET 1
(water-treated rats)

Name______________________________

Date______________________________

Fill in the information below:

species__________ route__________ % sol./susp.__________
dose_____________ drug____________ vehicle______________

Weigh the first animal and record the weight, calculate and administer the appropriate dose, and record the volume administered and the time. Repeat this procedure for all animals in the experiment.

Animal	1	2	3	4	5	6	7	8	9	10
Weight of animal (g)										
Volume administered (cc)										
Time dosed										

Observe the group as a whole (see Appendix D for definitions and observation techniques). Record the degree of response (0 = normal, + = slight, ++ = moderate, +++ = marked) after each item for the group. Take periodic observations until the group has returned to normal.

Time group observed					
Number dead/number dosed					
1. Activity increase					
2. Activity decrease					
3. Ataxia					
4. Coma					
5. Convulsion(s)					
6. Cyanosis					
7. Defecation, excessive					
8. Heart rate increase					
9. Heart rate decrease					
10. Irritability increase					
11. Irritability decrease					
12. Lacrimation increase					
13. Lacrimation decrease					
14. Muscle tone increase					
15. Muscle tone decrease					
16. Pain reflex loss					
17. Palpebral opening increase					
18. Palpebral opening decrease					
19. Piloerection					
20. Respiratory rate increase					
21. Respiratory rate decrease					
22. Righting reflex loss					
23. Salivation increase					
24. Startle response increase					
25. Startle response decrease					
26. Straub response					
27. Touch response increase					
28. Touch response decrease					
29. Tremors, twitches, jerks					
30. Urination increase					
31. Vasoconstriction increase					
32. Vasodilation increase					
33. Vocalization increase					

EXP. 6: OBSERVATION SHEET 2
(drug-treated rats)

Name____________________________

Date____________________________

Fill in the information below:

species__________ route__________ % sol./susp.__________
dose____________ drug___________ vehicle______________

Weigh the first animal and record the weight, calculate and administer the appropriate dose, and record the volume administered and the time. Repeat this procedure for all animals in the experiment.

Animal	1	2	3	4	5	6	7	8	9	10
Weight of animal (g)										
Volume administered (cc)										
Time dosed										

Observe the group as a whole (see Appendix D for definitions and observation techniques). Record the degree of response (0 = normal, + = slight, ++ = moderate, +++ = marked) after each item for the group. Take periodic observations until the group has returned to normal.

Time group observed					
Number dead/number dosed					
1. Activity increase					
2. Activity decrease					
3. Ataxia					
4. Coma					
5. Convulsion(s)					
6. Cyanosis					
7. Defecation, excessive					
8. Heart rate increase					
9. Heart rate decrease					
10. Irritability increase					
11. Irritability decrease					
12. Lacrimation increase					
13. Lacrimation decrease					
14. Muscle tone increase					
15. Muscle tone decrease					
16. Pain reflex loss					
17. Palpebral opening increase					
18. Palpebral opening decrease					
19. Piloerection					
20. Respiratory rate increase					
21. Respiratory rate decrease					
22. Righting reflex loss					
23. Salivation increase					
24. Startle response increase					
25. Startle response decrease					
26. Straub response					
27. Touch response increase					
28. Touch response decrease					
29. Tremors, twitches, jerks					
30. Urination increase					
31. Vasoconstriction increase					
32. Vasodilation increase					
33. Vocalization increase					

EXPERIMENT 7

Intraperitoneal and Subcutaneous Administration of Xylazine in the Rat

INTRODUCTION

This experiment acquaints the student with giving intraperitoneal (IP) and subcutaneous (SC) injections.

In this experiment, five rats are injected intraperitoneally with water, five rats are injected intraperitoneally with xylazine (Rompun), five rats are injected subcutaneously with water, and five rats are injected subcutaneously with xylazine. The four groups are then compared to determine the effects of xylazine in rats and the variation in absorption times of the two methods of injection.

MATERIALS AND EQUIPMENT (per student)

xylazine (Rompun), 20 mg/cc (20 cc vial per 20 students)
graduated cylinder, 10 cc
sterile beaker with 50 cc sterile distilled water
needle, 25 gauge, 1/4 in.
rat scale and balance weights
2 syringes, 1 cc
20 rats, each approximately 100 g
3 empty cages
needle, 25 gauge, 1/2 in.

PROCEDURE

Dose (IP and SC)

The anesthetic dose of xylazine is 1 mg/lb in rats. Because they are so small, rats are weighed in the metric system. Since all calculations must be made in the same system, the dose must be changed to mg/g (1 lb = 454 g--memorize this conversion because you will need it later).

$$\frac{1 \text{ mg xylazine}}{1 \text{ lb rat}} = \frac{1 \text{ mg xylazine}}{454 \text{ g rat}}$$

To calculate what a 100 gram rat will receive,

$$\frac{x \text{ mg xylazine}}{100 \text{ g rat}} = \frac{1 \text{ mg xylazine}}{454 \text{ g rat}}$$

$$(x)(454 \text{ g}) = (100 \text{ g})(1 \text{ mg})$$

$$x = 0.22 \text{ mg/100 g rat}$$

Volume (IP and SC)

The concentration of xylazine in the vial is 20 mg/1 cc. On the student report (number 3), calculate the volume to give a 100 g rat (answer: 0.011 cc).

Since you cannot accurately measure this volume in your syringe, you must dilute the concentration and give a larger volume.

Dilution

Place the 25 gauge, 1/4 in. needle on the syringe. Take 1 cc from the xylazine vial and place it carefully into the 10 cc graduated cylinder and q.s. the xylazine with sterile distilled water to 10 cc. You are diluting the 20 mg/1 cc to 20 mg/10 cc = 2 mg/1 cc).

Remember, you must correct the volume for any animal weighing more or less than 100 g (e.g., a 110 g rat would receive 0.12 cc diluted xylazine.)

IP injection

Check your scale for accuracy and adjust if necessary (see Exp. 1).

The IP control animals will be dosed first. Weigh the first rat and record the weight on observation sheet 1. Calculate the volume to give according to this weight and record it on the observation sheet.

The same syringe that was used for the dilution may be used again, but it must be rinsed carefully with sterile distilled water. To rinse the syringe, withdraw about 1 cc sterile water into the syringe and then express this water into the sink. Do this twice to be sure the syringe is fully rinsed. Then fill your syringe with the correct volume of sterile distilled water for administration and lay it aside.

Have your lab partner restrain your rat (see Exp. 1) and turn it approximately 120° (so its abdomen is facing you).

Pick up your syringe in your right hand and hold it as you did for an IM injection (Exp. II). With your left hand hold the rat's right hind foot. Insert your needle about 1/8 in. at a 45° angle approximately 1/2 in. caudal to (below) its umbilicus (Fig. 7.1).

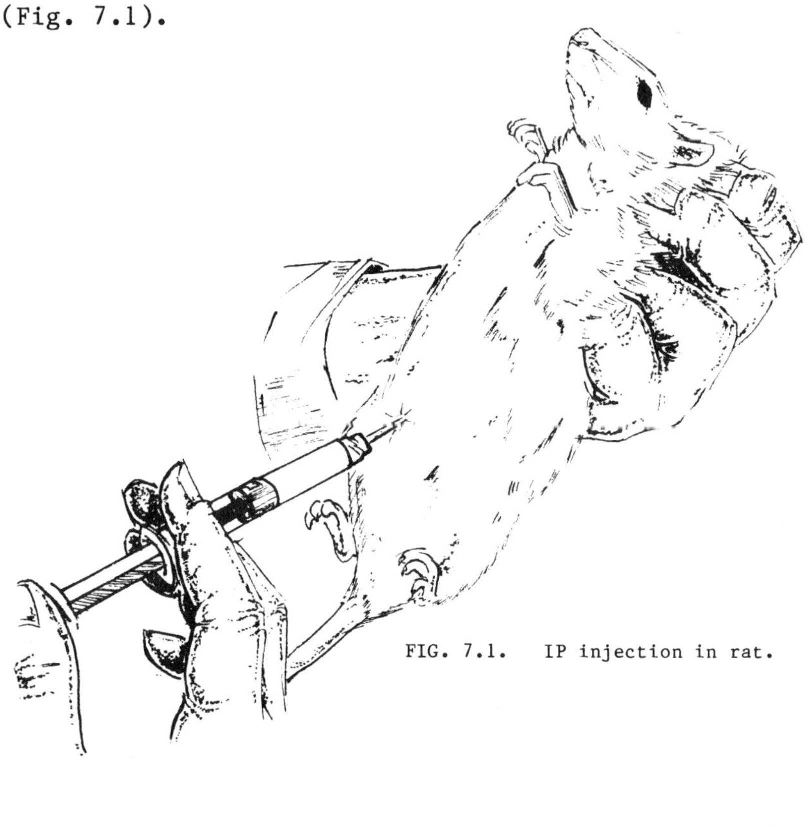

FIG. 7.1. IP injection in rat.

Inject the sterile distilled water and withdraw the syringe. Place the animal in the empty cage. Repeat the procedure with four additional rats (the control animals). Then, using diluted xylazine, repeat this procedure with 5 additional rats. After dosing these two groups of rats, reverse roles with your partner and repeat this procedure with your partner's two groups of rats.

SC Injection

Students will work individually giving SC in injections. Place the 1/2 in. needle on the second syringe.

Weigh the first rat, record the weight, and calculate and record the volume to give, as for the IP injection. Fill the syringe with sterile distilled water and lay it aside.

Restrain your rat in your left hand and pick up the syringe in your right hand. With the rat facing the needle, insert the needle through the skin in the neck region just in front of and below your fingers (Fig. 7.2). Inject the fluid and withdraw the needle. Place the animal in the third empty cage. Repeat the procedure with four additional rats (the control animals). Then dose the 5 remaining rats with the dose of xylazine in the same manner.

FIG. 7.2. SC injection in rat.

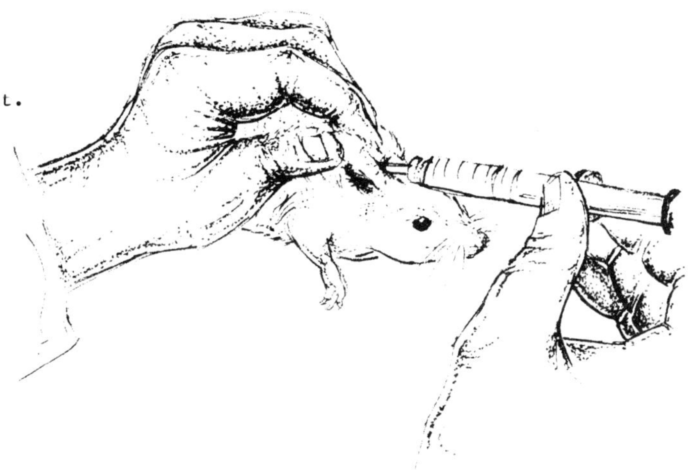

Results

Observe all four groups of animals. Be sure to note the time of onset of symptoms as well as the duration of symptoms. Record the observations on the four observation sheets.

Complete the student report sheet.

EXP. 7: STUDENT REPORT

Name______________________________

Date______________________________

1. What type of drug is xylazine? (Read the insert that accompanies the vial.)

2. Check the side effects of xylazine printed on the drug insert. What is emesis? Will emesis occur in rats?

3. Calculate the volume of xylazine in the vial to give a 100 g rat (show your work).

4. Was there a difference in the time of onset of symptoms between the IP- and SC-dosed animals?________________If so, why? If not, why not?

5. For anesthesia, Rompun is recommended at a dose of 2 ml/20 lb and a concentration of 10 mg/ml. Calculate the volume a 90 g rat should receive (show your work).

6. How many milligrams equal 1 lb?____________

EXP. 7: OBSERVATION SHEET 1
(water-treated rats, IP)

Name_______________________________

Date_______________________________

Fill in the information below:

species__________ route__________ % sol./susp.__________
dose__________ drug__________ vehicle______________

Weigh the first animal and record the weight, calculate and administer the appropriate dose, and record the volume administered and the time. Repeat this procedure for all animals in the experiment.

Animal	1	2	3	4	5
Weight of animal (g)					
Volume administered (cc)					
Time dosed					

Observe the group as a whole (see Appendix D for definitions and observation techniques). Record the degree of response (0 = normal, + = slight, ++ = moderate, +++ = marked) after each item for the group. Take periodic observations until the group has returned to normal.

Time group observed					
Number dead/number dosed					
1. Activity increase					
2. Activity decrease					
3. Ataxia					
4. Coma					
5. Convulsion(s)					
6. Cyanosis					
7. Defecation, excessive					
8. Heart rate increase					
9. Heart rate decrease					
10. Irritability increase					
11. Irritability decrease					
12. Lacrimation increase					
13. Lacrimation decrease					
14. Muscle tone increase					
15. Muscle tone decrease					
16. Pain reflex loss					
17. Palpebral opening increase					
18. Palpebral opening decrease					
19. Piloerection					
20. Respiratory rate increase					
21. Respiratory rate decrease					
22. Righting reflex loss					
23. Salivation increase					
24. Startle response increase					
25. Startle response decrease					
26. Straub response					
27. Touch response increase					
28. Touch response decrease					
29. Tremors, twitches, jerks					
30. Urination increase					
31. Vasoconstriction increase					
32. Vasodilation increase					
33. Vocalization increase					

EXP. 7: OBSERVATION SHEET 2
(drug-treated rats, IP)

Name__________________________

Date__________________________

Fill in the information below:

species__________ route__________ % sol./susp.__________
dose__________ drug__________ vehicle__________

Weigh the first animal and record the weight, calculate and administer the appropriate dose, and record the volume administered and the time. Repeat this procedure for all animals in the experiment.

Animal	1	2	3	4	5
Weight of animal (g)					
Volume administered (cc)					
Time dosed					

Observe the group as a whole (see Appendix D for definitions and observation techniques). Record the degree of response (0 = normal, + = slight, ++ = moderate, +++ = marked) after each item for the group. Take periodic observations until the group has returned to normal.

Time group observed					
Number dead/number dosed					
1. Activity increase					
2. Activity decrease					
3. Ataxia					
4. Coma					
5. Convulsion(s)					
6. Cyanosis					
7. Defecation, excessive					
8. Heart rate increase					
9. Heart rate decrease					
10. Irritability increase					
11. Irritability decrease					
12. Lacrimation increase					
13. Lacrimation decrease					
14. Muscle tone increase					
15. Muscle tone decrease					
16. Pain reflex loss					
17. Palpebral opening increase					
18. Palpebral opening decrease					
19. Piloerection					
20. Respiratory rate increase					
21. Respiratory rate decrease					
22. Righting reflex loss					
23. Salivation increase					
24. Startle response increase					
25. Startle response decrease					
26. Straub response					
27. Touch response increase					
28. Touch response decrease					
29. Tremors, twitches, jerks					
30. Urination increase					
31. Vasoconstriction increase					
32. Vasodilation increase					
33. Vocalization increase					

EXP. 7: OBSERVATION SHEET 3
(water-treated rats, SC)

Name____________________

Date____________________

Fill in the information below:

species________ route________ % sol./susp.________
dose________ drug________ vehicle________

Weigh the first animal and record the weight, calculate and administer the appropriate dose, and record the volume administered and the time. Repeat this procedure for all animals in the experiment.

Animal	1	2	3	4	5
Weight of animal (g)					
Volume administered (cc)					
Time dosed					

Observe the group as a whole (see Appendix D for definitions and observation techniques). Record the degree of response (0 = normal, + = slight, ++ = moderate, +++ = marked) after each item for the group. Take periodic observations until the group has returned to normal.

Time group observed					
Number dead/number dosed					
1. Activity increase					
2. Activity decrease					
3. Ataxia					
4. Coma					
5. Convulsion(s)					
6. Cyanosis					
7. Defecation, excessive					
8. Heart rate increase					
9. Heart rate decrease					
10. Irritability increase					
11. Irritability decrease					
12. Lacrimation increase					
13. Lacrimation decrease					
14. Muscle tone increase					
15. Muscle tone decrease					
16. Pain reflex loss					
17. Palpebral opening increase					
18. Palpebral opening decrease					
19. Piloerection					
20. Respiratory rate increase					
21. Respiratory rate decrease					
22. Righting reflex loss					
23. Salivation increase					
24. Startle response increase					
25. Startle response decrease					
26. Straub response					
27. Touch response increase					
28. Touch response decrease					
29. Tremors, twitches, jerks					
30. Urination increase					
31. Vasoconstriction increase					
32. Vasodilation increase					
33. Vocalization increase					

EXP. 7: OBSERVATION SHEET 4
(drug-treated rats, SC)

Name____________________________

Date____________________________

Fill in the information below:

species__________ route__________ % sol./susp.__________
dose__________ drug__________ vehicle__________

Weigh the first animal and record the weight, calculate and administer the appropriate dose, and record the volume administered and the time. Repeat this procedure for all animals in the experiment.

Animal	1	2	3	4	5
Weight of animal (g)	____	____	____	____	____
Volume administered (cc)	____	____	____	____	____
Time dosed	____	____	____	____	____

Observe the group as a whole (see Appendix D for definitions and observation techniques). Record the degree of response (0 = normal, + = slight, ++ = moderate, +++ = marked) after each item for the group. Take periodic observations until the group has returned to normal.

Time group observed					
Number dead/number dosed					
1. Activity increase					
2. Activity decrease					
3. Ataxia					
4. Coma					
5. Convulsion(s)					
6. Cyanosis					
7. Defecation, excessive					
8. Heart rate increase					
9. Heart rate decrease					
10. Irritability increase					
11. Irritability decrease					
12. Lacrimation increase					
13. Lacrimation decrease					
14. Muscle tone increase					
15. Muscle tone decrease					
16. Pain reflex loss					
17. Palpebral opening increase					
18. Palpebral opening decrease					
19. Piloerection					
20. Respiratory rate increase					
21. Respiratory rate decrease					
22. Righting reflex loss					
23. Salivation increase					
24. Startle response increase					
25. Startle response decrease					
26. Straub response					
27. Touch response increase					
28. Touch response decrease					
29. Tremors, twitches, jerks					
30. Urination increase					
31. Vasoconstriction increase					
32. Vasodilation increase					
33. Vocalization increase					

EXPERIMENT 8

Common Procedures and Techniques in the Dog

EXPERIMENT 8A. Determination of Heart Rate, Pulse Rate, and Temperature in the Dog

---

INTRODUCTION

Experiment 8A acquaints the student with taking the heart and pulse rate, the respiratory rate, and the rectal temperature of a dog.

Students work in pairs throughout this experiment. One student restrains the dog gently while the partner performs a particular part of the procedure. Then the partners reverse roles and repeat the procedure.

MATERIALS AND EQUIPMENT (per 2 students)

- stethoscope
- dog
- rectal thermometer
- lubricating agent (soap, Vaseline, etc.) (1 jar per class)
- 4 cotton balls or tissues
- thermometer holder with alcohol
- 50 cc bottle of alcohol, isopropyl, 70% (per 20 students) (if thermometer holder not used)

PROCEDURE

Handling and Lifting

Dogs show a great variation in temperament and personality. Before handling a dog, speak to it quietly in a friendly voice. Always approach it slowly and quietly. Observe carefully how the animal reacts (i.e., friendly, nervous, scared) and handle the animal accordingly. Most dogs are friendly and allow themselves to be handled easily.

To lift a small dog onto the examination or lab table, bend your knees and put your left hand under its chest between its front legs. Place your right hand under and around its neck and lift up.

To lift a large dog onto the table, bend your knees and put one arm around its chest and the other arm around its rear thighs. Lift it in both arms with its side against you.

Mild Restraint

Dogs may be restrained by several different methods. The method used depends on the animal's temperament, the purpose of the restraint, or the method most familiar to or liked by a supervisor.

Mild restraint is necessary to determine an accurate heart rate, pulse, and respiratory rate. Once an animal starts to struggle or becomes excited, these parameters will increase.

Have your partner lift the dog quietly onto the table and hold it by its collar. Usually this is all the restraint needed to keep the animal on the table.

Heart Rate Determination

The heart rate is the number of heartbeats per minute. Each typical heartbeat has two sounds, lubb-dubb, followed by a rest. The first sound (lubb) is low and dull and occurs during ventricular systole. It is caused by the contraction of the ventricle muscle and the closing of the valves between the atria and ventricles. The second sound (dubb) is shorter, louder, and higher pitched. It is caused by the semilunar valves closing.

Abnormal sounds may indicate heart problems; for example, a soft swishing sound may indicate a murmur, which is blood leaking back through an imperfectly closed valve.

To take the dog's heart rate, place the stethoscope earpieces, angled toward your face, into your ears. Place the round head of the stethoscope on the dog's chest just behind the front limb.

Count the number of beats (lubb-dubbs) over a 15 sec period. Record this number and the heartbeats per minute (the number of beats per 15 sec multiplied by 4) on your student report sheet.

The normal heart rate of a dog is 62-130 beats per min. The rate varies in an individual because of changes in ambient and body temperature, stress, time of day, activity, etc.

Pulse Rate Determination

Each heartbeat causes a surge of blood, or pulse. The pulse rate, which may be felt in any superficial artery, measures the same as the heart rate. In the dog one of the best sites to feel the pulse is on the medial side of the thigh (groin area), approximately midway between the caudal and cranial parts of the thigh where the femoral artery lies. While your partner holds the dog in a standing or lying position on the table, use your index and middle fingers to palpate the artery. Count and record the number of pulses over a 15 sec period. Multiply that number by 4 and record the pulse rate per minute on the student report sheet.

Respiratory Rate Determination

Respiration (breathing) is the gaseous interchange between an organism and its environment. The respiratory rate is the number of respirations per minute. To determine the respiratory rate, have the dog lie quietly on the table or floor. Count the number of inspirations (or raising of the chest cavity) for 15 sec. Multiply this number by 4 to determine the respiratory rate per minute and record it on your student report sheet.

The normal respiratory rate for dogs is 10-30 breaths/min.

Temperature Determination

The thermometer is a useful tool for determining abnormal body conditions. If an animal is sick, the temperature should be taken every 4 hr during the day because some febrile (feverish) diseases may cause an intermittent rise and fall of temperature.

There are several types of thermometers. The rectal thermometer is generally used for animals.

To read the thermometer, hold the end opposite the silver mercury bulb so that you can read the numbers. Rotate the thermometer, turning the numbers downward toward the floor. Look for the silver mercury "ribbon" running along part of the length of the thermometer. The end of this ribbon marks the temperature reading. You will notice two sizes of ruler-type markings along the thermometer. The longer lines indicate the whole numbers of degrees (e.g., 95°, 96°). The shorter lines indicate 2/10 of a degree (e.g., 95.2°, 95.4°) (Fig. 8.1). An arrow points to 98.6°F., which is considered normal in the human.

FIG. 8.1. Rectal thermometer reading 99.2°F.

Before taking the temperature, the mercury must be below this "normal" mark (generally below 98°F). To shake down the thermometer, hold the end opposite the bulb with your thumb and index finger. Make sure you are not near any objects (lab bench, chair, etc.) on which you could accidentally strike the thermometer and break it. Raise your hand holding the thermometer chest high and with a quick downward movement lower your hand and snap your wrist sharply. You may have to repeat this procedure several times to drive the mercury below 98°F.

Have your lab partner hold the dog in a standing position on the table. Generally mild restraint with your arm around its neck will suffice, but if the dog is very jumpy or you suspect it might bite, use moderate restraint and put a gauze muzzle on it (see Exp. 8B for a description of moderate restraint).

To take the temperature, moisten or lubricate the thermometer with soap, Vaseline, or some other suitable agent so it will slide easily into the rectum. If you gently exert slight pressure with the thermometer bulb against the anus, the animal generally will relax the anus muscles and the thermometer will slide into the rectum. Insert the bulb about 1 in. into the rectum and leave it in place about 3 min. Then withdraw it and wipe the end with a piece of cotton or tissue. Read the temperature and have your partner check your reading. Record the temperature on your student report sheet.

When finished with the thermometer, replace it in its jar of antiseptic solution, or disinfect it using a piece of cotton soaked in 70% alcohol and then return it to its holder.

The normal temperature of a dog is 102° $\pm$ 1°F.

Complete your student report sheet.

EXP. 8A: STUDENT REPORT

Name______________________________

Date______________________________

1. Record the following information about the dog:

    a. Name, number, or description--

    b. Sex________________________.

    c. Heart rate________________.

    d. Pulse rate________________.

    e. Respiratory rate__________.

    f. Rectal temperature________.

2. If any of these four measured parameters are abnormal, suggest a reasonable explanation.

EXPERIMENT 8B. Intravenous Administration of Xylazine in the Dog

INTRODUCTION

This experiment acquaints the student with applying moderate restraint to the dog, giving intravenous (IV) injections in the dog, and the effects of xylazine in the dog.

Xylazine is a nonnarcotic compound that causes muscle relaxation, sedation, analgesia, and reduced respiratory and heart rates. It is used in dogs and cats to produce sedation and/or a short period of analgesia (e.g., for dental procedures, minor surgery, or as a tranquilizer).

Students work in pairs. One student loads the syringe, makes the gauze muzzle and places it on the dog (if necessary), shaves the foreleg, and administers the xylazine. The lab partner weighs the dog, applies mild restraint, and then occludes the vein. Both students should independently calculate the dose and observe and record the results. In the following experiment (Exp. 8C), the partners reverse roles.

MATERIALS AND EQUIPMENT (per 2 students)

- materials and equipment listed in Exp. 8A
- human scale (per 10 students)
- syringe, 10 cc
- needle, 20 gauge, 1 in.
- xylazine (Rompun), 20 mg/ml injectable, 20 cc/vial (1-3 vials per 20 students, depending on the weights of the dogs)
- gauze bandage, 3-4 ft length (optional)
- surgical clippers (per 4 students)
- 4 cotton balls
- 50 cc bottle of alcohol, isopropyl, 70% (per 20 students)

PROCEDURE

Weighing

Check your scale for accuracy and adjust if necessary (see Exp. 1).

Weigh yourself first. Then pick up the dog gently and weigh both yourself and the dog. Subtract your weight from the total weight. The difference will be the dog's weight. Record the dog's weight on the observation sheet.

Dose

The dose is 0.5 mg/lb, IV. On the student report (number 2), calculate how many milligrams to give a 20 lb dog (answer: 10 mg).

On the student report (number 3), calculate how many milligrams to give your dog according to its weight.

Volume

The concentration in the vial is 20 mg/ml. On your student report (number 4), calculate the volume to be given to a 20 lb dog (answer: 0.5 ml).

On your student report (number 5), calculate the volume to give your dog and record it on the observation sheet. Have your instructor check your calculations.

Attach the needle to the 10 cc syringe and fill to the desired volume with xylazine. Make sure all trapped air is expelled. (If you are unsure of how to do this, see Exp. 1.) Lay the filled syringe aside.

Moderate Restraint

Have your lab partner lift the dog gently onto the examination or lab table. If you know or suspect the animal will bite, make a gauze muzzle and place it around its mouth.

To make a gauze muzzle, take a 3 or 4 ft length of gauze and fold it in half lengthwise. Place a loose single knot in the middle, forming a noose that is about 3 in. in diameter (Fig. 8.2). Hold the noose with the knot downward. Then, standing near the side of the dog, place the noose over its nose and tie the ends quickly in a slip knot behind its ears (Fig. 8.3).

FIG. 8.2. Single knot placed in double length of gauze used to make muzzle.

FIG. 8.3. Gauze muzzle placed around mouth of dog.

To moderately restrain the dog, your partner should stand on the dog's left side (turn the dog around if necessary) against the dog's body. Holding the dog's muzzle gently but firmly in the left hand with the fingers under the jaw and the thumb over the snout (Fig. 8.4), your partner should place the right arm over the dog's shoulder and the right palm behind and around its right elbow. Then your partner can extend the leg for shaving or IV administration by exerting slight pressure against the elbow.

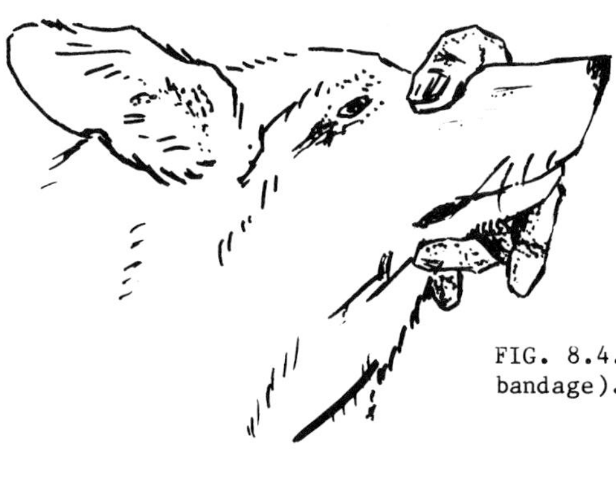

FIG. 8.4. Holding dog's muzzle (shown without gauze bandage).

On the dog's leg extended by your partner, clip the hair that lies over the cephalic vein area of the upper anterior foreleg. The clipped area should be about 1 by 2 in. long, beginning about 1/2 in. below the chest and extending down the leg.

IV Injection

Wipe the area with a cotton ball saturated with 70% alcohol to disinfect the skin.

With the right hand still holding the dog's leg, your partner applies pressure against the anterior foreleg with the right thumb, occluding the cephalic vein. This makes the vein raise or balloon up. Your partner keeps the vein occluded until told to release the pressure.

Then pick up the loaded syringe in the right hand and place your left hand under the shaved area of the extended leg. Extend your fingers around the leg and push the left thumb slightly against the raised vein so that when the needle is inserted the vein won't roll away from the needle (Fig. 8.5).

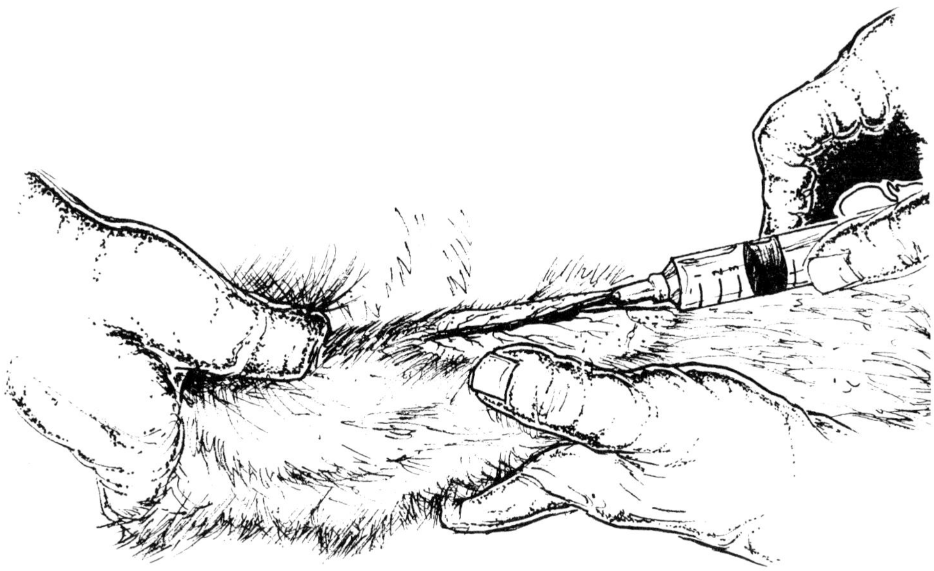

FIG. 8.5. Inserting needle into cephalic vein of dog's leg.

Insert the needle at about a 10° angle through the skin and into the vein, all in one motion. Extend your left thumb and index finger over the syringe to steady it. Pull back slightly on the plunger. If no blood flows into the syringe, the needle is not in the vein. Withdraw the needle about 1/4 in. and apply negative pressure against the plunger again to see if blood flows into the syringe. Sometimes the needle may have been pushed completely through the vein.

Once the needle is in the vein, have your lab partner loosen the pressure on the vein. Then inject the xylazine slowly into the vein. If the fluid does not flow easily and if a swelling appears in the leg, the needle is no longer in the vein. Withdraw the needle about 1/8 in. and try again. If there is no improvement in the flow or if the area continues to swell, discontinue the injection, place your thumb that was alongside the vein over the injection site, withdraw the needle, and assert pressure on the vein for a minute or so. Try again in the other leg.

Once the fluid is injected, place your thumb that was alongside the vein over the injection site and assert slight pressure. Remove the needle and assert more pressure to keep the vein from seeping. Apply pressure for about 1 min.

The animal will enter a sleeplike state within 3-5 min. If you applied a gauze muzzle, remove it now.

Observe the dog and record your observations on the observation sheet and your student report sheet (numbers 6 and 7).

EXP. 8B: STUDENT REPORT

Name_______________________

Date_______________________

1. What type of drug is xylazine and what are its effects?

2. Calculate the milligrams to give a 20 lb dog (show your work).

3. Calculate the milligrams to give your dog, according to its weight (show your work).

4. Calculate the volume to give a 20 lb dog (show your work).

5. Calculate the volume to give your dog, according to its weight (show your work).

6. Record the following information about the vital signs 5 minutes after administration of xylazine:

   a. Name, number, or description--

   b. Sex________________________.

   c. Heart rate_________________.

   d. Pulse rate_________________.

   e. Respiratory rate__________.

   f. Rectal temperature________.

7. If any of the four measured parameters are abnormal, suggest a reasonable explanation.

EXP. 8B: OBSERVATION SHEET

Name____________________________

Date____________________________

Fill in the information below:

species__________ route__________ % sol./susp.__________

dose__________ drug__________ vehicle__________

Weigh the animal and record the weight, calculate and administer the appropriate dose, and record the volume administered and the time.

Weight of animal (lb) __________

Volume administered (cc) __________

Time dosed __________

Observe the animal (see Appendix D for definitions and observation techniques). Record the degree of response (0 = normal, + = slight, ++ = moderate, +++ = marked) after each item. Take periodic observations until the animal has returned to normal.

Time observed					
Number dead/number dosed					
1. Activity increase					
2. Activity decrease					
3. Ataxia					
4. Coma					
5. Convulsion(s)					
6. Corneal reflex loss					
7. Cyanosis					
8. Defecation, excessive					
9. Heart rate increase					
10. Heart rate decrease					
11. Irritability increase					
12. Irritability decrease					
13. Lacrimation increase					
14. Lacrimation decrease					
15. Muscle tone increase					
16. Muscle tone decrease					
17. Pain reflex loss					
18. Palpebral opening increase					
19. Palpebral opening decrease					
20. Piloerection					
21. Respiratory rate increase					
22. Respiratory rate decrease					
23. Righting reflex loss					
24. Salivation increase					
25. Startle response increase					
26. Startle response decrease					
27. Straub response					
28. Touch response increase					
29. Touch response decrease					
30. Tremors, twitches, jerks					
31. Urination increase					
32. Vasoconstriction increase					
33. Vasodilation increase					
34. Vocalization increase					

EXPERIMENT 8C. Antagonism of Xylazine with Doxapram Hydrochloride Administered Intravenously in the Dog

INTRODUCTION

This experiment acquaints the student with the stimulatory effects of doxapram hydrochloride (Dopram-V), a potent respiratory stimulant in the dog.

In Exp. 8B the dog received xylazine, 0.5 mg/lb, IV, which produced anesthesia, sedation, and muscle relaxation and reduced the respiratory rate and heart rate. In this experiment, doxapram hydrochloride is injected into the dog while it is still experiencing these effects. The dog is then observed to determine the antagonistic effects of doxapram hydrochloride.

MATERIALS AND EQUIPMENT (per 2 students)

- materials and equipment listed in Exp. 8A
- dog injected with xylazine
- syringe, 10 cc
- needle, 20 gauge, 1 in.
- doxapram hydrochloride (Dopram-V), injectable, 20 cc/vial (1-2 vials per 1-4 students, depending on weights of dogs)

PROCEDURE

Dose and Volume

The dose is 4 mg/lb, IV. On your student report (number 2), calculate the milligrams and volume of Dopram-V to give the dog (a 20 lb dog would receive 4 cc). Record these on the observation sheet. Have your instructor check your calculations.

IV Injection

(Laboratory partners reverse roles for Exp. 8C.) Clip the dog's left leg and swab the area with alcohol while your partner holds the dog's left leg extended. Then the cephalic vein should be occluded by your partner so the vein will raise (see Exp. 8B).

Load the syringe to the correct volume and make sure no air bubbles are trapped in the syringe. Then insert the needle into the vein (as in Exp. 8B). Exert negative pressure on the plunger to make sure you are in the vein (blood will flow into the syringe).

Once you are sure the needle is in the vein, have your lab partner release the thumb pressure on the vein and then inject the Dopram-V slowly. Apply thumb pressure over the injection site and withdraw the needle.

Observe the dog and record your observations on the observation sheet and student report sheet (numbers 6 and 7).

Complete your student report sheet.

EXP. 8C: STUDENT REPORT

Name____________________________

Date____________________________

1. What type of drug is Dopram-V and what are its effects?

2. Calculate the milligrams and volume to give your dog (show your work).

3. For a 2.5 mg/lb dose of xylazine at a concentration of 30 mg/ml, calculate the volume to give a 25 lb dog (show your work).

4. For a 3 mg/lb dose of Dopram-V at a concentration of 15 mg/ml, calculate the volume to give a 33 lb dog (show your work).

5. Does Dopram-V work as a respiratory stimulant in dogs with the following anesthesias? (Read the Dopram-V package insert.)

   a. Pentobarbital ___________

   b. Thiopental ___________

   c. Halothane ___________

   d. Morphine ___________

6. Record the following information about the vital signs 5 minutes after administration of Dopram-V:

   a. Name, number, or description--

   b. Sex________________________.

   c. Heart rate________________.

   d. Pulse rate________________.

   e. Respiratory rate__________.

   f. Rectal temperature________.

7. If any of the four measured parameters differ from those recorded in Exp. 8A or 8B, suggest reasonable explanations.

EXP. 8C: OBSERVATION SHEET

Name______________________________

Date______________________________

Fill in the information below:

species__________ route__________ % sol./susp.__________
dose__________ drug__________ vehicle__________

Weigh the animal and record the weight, calculate and administer the appropriate dose, and record the volume administered and the time.

Weight of animal (lb) ____________
Volume administered (cc) ____________
Time dosed ____________

Observe the animal (see Appendix D for definitions and observation techniques). Record the degree of response (0 = normal, + = slight, ++ = moderate, +++ = marked) after each item. Take periodic observations until the animal has returned to normal.

Time observed					
Number dead/number dosed					
1. Activity increase					
2. Activity decrease					
3. Ataxia					
4. Coma					
5. Convulsion(s)					
6. Corneal reflex loss					
7. Cyanosis					
8. Defecation, excessive					
9. Heart rate increase					
10. Heart rate decrease					
11. Irritability increase					
12. Irritability decrease					
13. Lacrimation increase					
14. Lacrimation decrease					
15. Muscle tone increase					
16. Muscle tone decrease					
17. Pain reflex loss					
18. Palpebral opening increase					
19. Palpebral opening decrease					
20. Piloerection					
21. Respiratory rate increase					
22. Respiratory rate decrease					
23. Righting reflex loss					
24. Salivation increase					
25. Startle response increase					
26. Startle response decrease					
27. Straub response					
28. Touch response increase					
29. Touch response decrease					
30. Tremors, twitches, jerks					
31. Urination increase					
32. Vasoconstriction increase					
33. Vasodilation increase					
34. Vocalization increase					

EXPERIMENT 9

Common Procedures and Techniques in the Cat

EXPERIMENT 9A. Determination of Heart Rate, Respiratory Rate, and Temperature in the Cat

INTRODUCTION

This experiment acquaints the student with determining the heart and respiratory rates and the rectal temperature of the cat.

Students work in pairs. One student performs the procedure while the lab partner holds the cat, using mild restraint. After each procedure the students reverse roles and repeat the procedure.

MATERIALS AND EQUIPMENT (per 2 students)

- cat
- stethoscope
- rectal thermometer
- lubricating agent (soap, Vaseline, etc.) (1 jar per class)
- 4 cotton balls or tissues
- thermometer holder with alcohol (optional)
- 50 cc bottle of alcohol, isopropyl, 70% (per 20 students) (if thermometer holder not used)

PROCEDURE

Mild Restraint

Place the cat quietly and gently on the examination or laboratory table. Cats do not like to be restrained. Since any emotional conflict will alter vital signs dramatically, hold the cat loosely around the neck.

Heart Rate Determination

Take the cat's heart rate with the stethoscope in the same manner as for the dog (Exp. 8A). Record the heart rate on the student report sheet.

The normal heart rate for cats is 110-130 beats/min.

Reverse roles with your partner and repeat the procedure.

Respiratory Rate Determination

Take the respiratory rate while the cat is lying quietly and record it on the student report sheet.

The normal respiratory rate for the cat is 25-35 breaths/min.

Reverse roles with your partner and repeat the procedure.

Temperature Determination

Check the temperature on the thermometer and shake down the mercury if necessary. Lubricate the thermometer and, if the cat is gentle and friendly and does not seem to resist the loose restraint, gently insert the thermometer in the rectum. Hold it gently in place for 3 min. Withdraw and read the thermometer and record the temperature on the student report.

Moderate Restraint

If it is necessary to use moderate restraint, have your lab partner position the cat standing or lying with all four feet squarely underneath it. This prevents the cat from scratching. Then your lab partner places the left hand over the cat's shoulders and the right hand over the cat's hips and gently but firmly presses and holds the cat down on the table in a crouching position (Fig. 9.1). You then can insert the thermometer and take the cat's temperature.

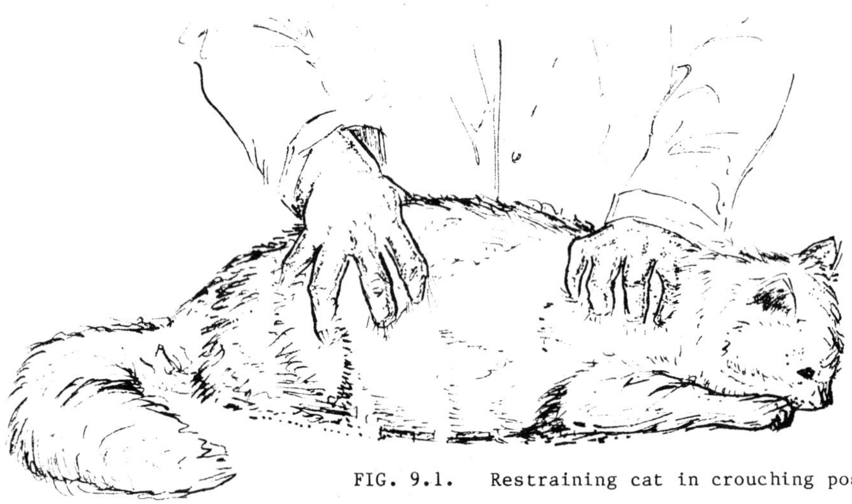

FIG. 9.1. Restraining cat in crouching position.

The normal temperature of the cat is 101.5° ± 1°F.

Reverse roles with your partner and repeat the procedure.

Complete your student report sheet.

EXP. 9A: STUDENT REPORT

Name________________________

Date________________________

1. Record the following information about your cat:

   a. Name, number, or description--

   b. Sex____________________

   c. Heart rate________________

   d. Respiratory rate__________

   e. Rectal temperature________

2. If any of these three measured parameters were abnormal, suggest a reasonable explanation.

EXPERIMENT 9B. Intramuscular Administration of Ketamine Hydrochloride in the Cat

INTRODUCTION

This experiment acquaints the student with giving intramuscular (IM) injections and with the effects of ketamine in the cat.

Ketamine is a dissociative anesthesia, one that causes the individual to feel dissociated from or unaware of its environment. It is used in veterinary practice for restraint or minor surgery that does not require muscle relaxation.

Students work in pairs. One student performs the procedure while the lab partner weighs and restrains the cat. Both students independently calculate the dose and observe and record the results.

MATERIALS AND EQUIPMENT (per 2 students)

human scale (per 10 students)
cat
syringe, 1 cc
needle, 23 gauge, 1/4 in.
ketamine hydrochloride, 100 mg/ml (0.2-2.0 cc per student, depending on the weights of the cats)
50 cc bottle of alcohol, isopropyl, 70% (per 20 students)

PROCEDURE

Weighing

Check your scale for accuracy and adjust if necessary (see Exp. 1).

The lab partner weighs the cat (see Exp. 8B). Record the weight on the observation sheet.

Dose

The dose is 5 mg/lb. On your student report (number 1), calculate the milligrams to give according to the cat's weight and record this number on the observation sheet.

Volume

Using 100 mg/ml concentration, on your student report (number 2), calculate the volume of ketamine to give the cat. Record the volume on the observation sheet.

Have your instructor check your dose and volume calculations.

Firm Restraint

The lab partner restrains the cat firmly. Grab the cat's scruff (loose skin on the back of the neck) securely with the right hand and the hind feet in the left hand with one finger between the legs. Then stretch the cat's body slightly on the table (Fig. 9.2).

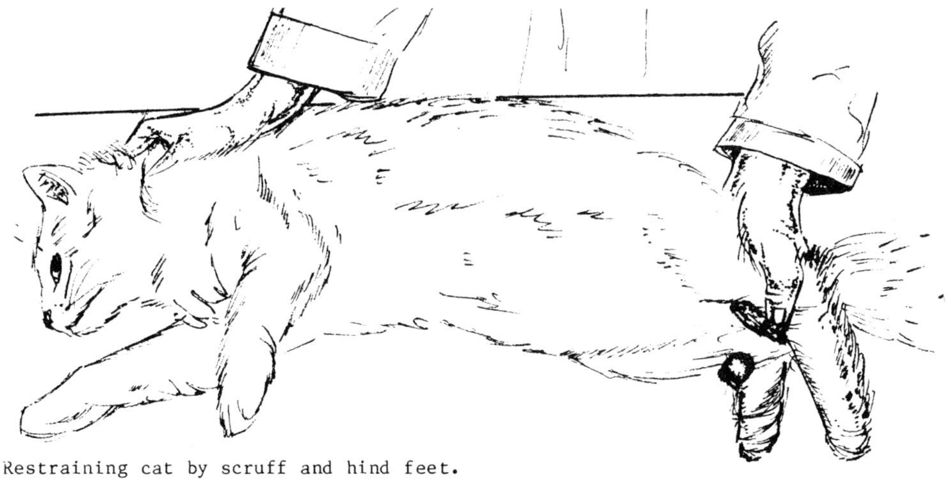

FIG. 9.2. Restraining cat by scruff and hind feet.

IM Injection

Attach the needle to the 1 cc syringe and fill the syringe to the correct volume with ketamine.

The IM injection is given in the gluteal region of the hind leg about halfway between the hip and the stifle joints. Moisten the cotton ball in alcohol and swab the area thoroughly so the skin becomes wet where the animal will be injected. Insert the needle perpendicularly through the skin and into the muscle (Fig. 9.3).

FIG. 9.3. IM injection in cat.

Quickly exert slight negative pressure on the plunger to make sure the needle is not in a blood vessel. If blood does not appear in the syringe, quickly inject the ketamine into the muscle.

Your lab partner then releases the cat from the firm restraint grip but keeps one hand on the animal's neck so it won't escape.

Record the onset and duration of sedation and the symptoms on the observation sheet.

Complete your student report sheet.

EXP. 9B: STUDENT REPORT

Name______________________________

Date______________________________

1. Calculate the milligrams to give the cat (show your work).

2. Calculate the volume to give (show your work).

3. Calculate the volume of ketamine to give a 7 lb cat for a dose of 4 mg/lb with a concentration of 80 mg/ml (show your work).

4. Name two other dissociative anesthetics. Look them up in _Veterinary Pharmacology and Therapeutics_, by Booth and McDonald (see the Bibliography).

EXP. 9B: OBSERVATION SHEET

Name ______________________________

Date ______________________________

Fill in the information below:

species __________ route __________ % sol./susp. __________
dose __________ drug __________ vehicle __________

Weigh the animal and record the weight, calculate and administer the appropriate dose, and record the volume administered and the time.

Weight of animal (lb) ____________
Volume administered (cc) ____________
Time dosed ____________

Observe the animal (see Appendix D for definitions and observation techniques). Record the degree of response (0 = normal, + = slight, ++ = moderate, +++ = marked) after each item. Take periodic observations until the animal has returned to normal.

Time observed					
Number dead/number dosed					
1. Activity increase					
2. Activity decrease					
3. Ataxia					
4. Coma					
5. Convulsion(s)					
6. Corneal reflex loss					
7. Cyanosis					
8. Defecation, excessive					
9. Heart rate increase					
10. Heart rate decrease					
11. Irritability increase					
12. Irritability decrease					
13. Lacrimation increase					
14. Lacrimation decrease					
15. Muscle tone increase					
16. Muscle tone decrease					
17. Pain reflex loss					
18. Palpebral opening increase					
19. Palpebral opening decrease					
20. Piloerection					
21. Respiratory rate increase					
22. Respiratory rate decrease					
23. Righting reflex loss					
24. Salivation increase					
25. Startle response increase					
26. Startle response decrease					
27. Straub response					
28. Touch response increase					
29. Touch response decrease					
30. Tremors, twitches, jerks					
31. Urination increase					
32. Vasoconstriction increase					
33. Vasodilation increase					
34. Vocalization increase					

EXPERIMENT 10

Oral Administration of Piperacetazine Hydrochloride and Subcutaneous Administration of Normal Saline in the Cat

INTRODUCTION

This experiment acquaints the student with administering piperacetazine hydrochloride (Psymod) orally in the cat. This drug is a psychotropic compound that produces a tranquilizing and sedative effect on dogs and cats with a minimum of drowsiness and ataxia. It is also used in veterinary medicine for the relief of pruritis, as a preanesthetic (parenteral route only), and as an antiemetic agent.

The same cat is then given normal saline by subcutaneous injection. Normal saline is used to restore both fluid and sodium chloride losses in animals. When given subcutaneously (usually to small animals that are not in critical condition due to these losses) the procedure is called hypodermoclysis.

Students will work in pairs. One student administers Psymod while the lab partner restrains the cat. Students switch roles for the administration of saline.

MATERIALS AND EQUIPMENT (per 2 students)

- human scale (per 10 students)
- cat
- piperacetazine hydrochloride (Psymod), 1 mg tablets (2-6 tablets per student, depending on the weights of the cats)
- pet piller (optional)
- sterile normal saline, 250 ml bottle (per 20 students)
- needle, 20 gauge, 1 in.
- syringe, 20 cc
- 50 cc bottle of alcohol, isopropyl, 70% (per 20 students)
- 4 cotton balls

PROCEDURE

<u>Weighing</u>

Check the scale for accuracy and adjust if necessary (see Exp. 1). Weigh the cat (see Exp. 8B) and record the weight on the observation sheet.

<u>Dose of Psymod</u>

The dose is 0.2 mg/kg. On your student report (number 3), calculate the milligrams to give the cat, according to its weight. Record this number on the observation sheet.

Volume

On your student report (number 4), calculate the number of tablets or the portion of a tablet to give to the cat. Record this number on the observation sheet.

Place the tablet(s) to be administered in a convenient position near the animal. If two tablets are to be administered, reverse roles with your partner for the administration of the second tablet.

Restraint and Administration, Oral

Have your lab partner use mild restraint (Exp. 9A) and place the cat in a sitting position, facing you on the edge of the table. One hand is placed on the cat's lower back so it can't back away from you. Then gently hold its head with your left hand so that your fingers are on one side of the head, your thumb is on the other side, and your palm is over the top of the head (Fig. 10.1).

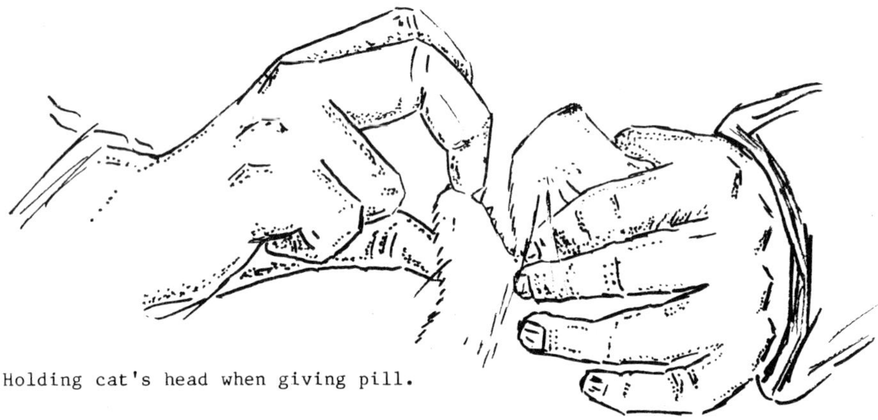

FIG. 10.1. Holding cat's head when giving pill.

Tilt the cat's nose toward the ceiling. Pick up the tablet with your right hand between index finger and thumb. With your middle finger on the cat's lower central incisor teeth, push downward on the lower jaw, forcing the jaw open. Push the tablet quickly over the back (base) of the cat's tongue. Quickly close its jaws and stroke the throat in a downward motion to help initiate swallowing (since the cat has a well developed gag reflex).

Some individuals prefer to use a commercial pilling device to deposit the tablet at the back of the tongue, particularly with a resistant animal.

Observe the animal and record symptoms on observation sheet 1.

Dose of Saline

The dose is 1.5 cc/kg. On your student report (number 6), calculate the volume to give your animal. Record this number on observation sheet 2.

Fill the 20 cc syringe with this volume and set it aside.

Restraint and Administration, SC

Have your lab partner restrain the cat, using either moderate or firm restraint (see Exps. 9A or 9B).

Swab the injection area with a cotton ball moistened with 70% alcohol (either the side of the neck or flank region).

With your left hand pick up a fold of skin at the injection site between your index finger and thumb. With your right hand pick up the filled syringe and insert the needle about 1/2 in. through the elevated skin (Fig. 10.2).

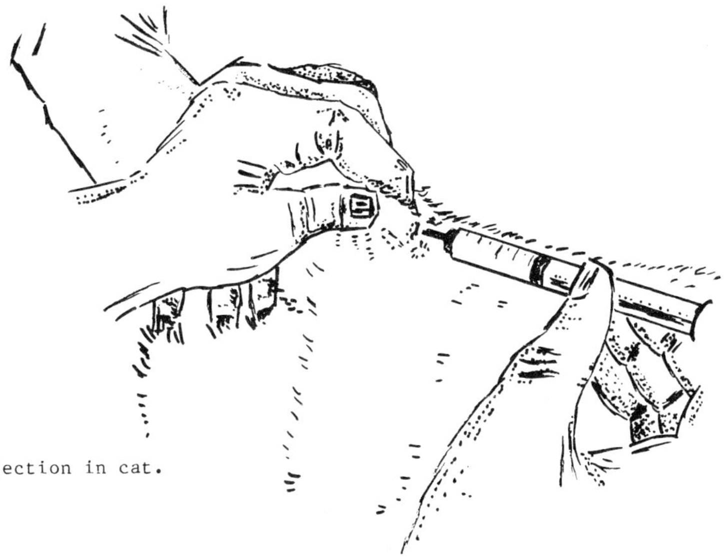

FIG. 10.2. SC injection in cat.

Exert slight negative pressure on the plunger to make sure the needle is not in a blood vessel and then inject the saline solution. The solution should flow easily into the area. If the solution does not flow readily, the needle might have penetrated the muscle or dermal layer of the skin. Try again if there is resistance to the flow.

After injecting the saline, withdraw the needle and gently massage the area to assist absorption of the fluid.

Observe the animal and record symptoms on observation sheet 2.

Complete your student report sheet.

EXP. 10: STUDENT REPORT

Name ______________________________

Date ______________________________

1. Define the following briefly:

   a. Psychotropic agent--

   b. Pruritis--

   c. Antiemetic--

   d. Hypodermoclysis--

2. Could Psymod be used for the following?

   a. Preventing car sickness__________

   b. Preventing nervous vomiting in cats____________

   c. Reducing fear of animals during thunderstorms____________

   d. Controlling anxiety in female cats during estrus____________

3. Calculate the milligrams of Psymod to give your cat (show your work).

4. Calculate the number of tablets or portion of a tablet to give your cat (show your work).

5. For a 0.5 mg/kg dose of Psymod, calculate how many tablets you should give a cat weighing 4 kg (show your work).

6. Calculate the volume of saline to give the 4 kg cat (show your work).

7. A cat has been hit by a car, has lost a considerable volume of blood, and is in critical condition:

   a. What route would you use to administer normal saline?

b. Why would you use this route?

c. Why would you administer normal saline?

EXP. 10: OBSERVATION SHEET 1
(drug-treated)

Name_______________

Date_______________

Fill in the information below:

species__________ route__________ % sol./susp.__________
dose__________ drug__________ vehicle__________

Weigh the animal and record the weight, calculate and administer the appropriate dose, and record the volume administered and the time.

Weight of animal (kg) __________
Volume administered (cc) __________
Time dosed __________

Observe the animal (see Appendix D for definitions and observation techniques). Record the degree of response (0 = normal, + = slight, ++ = moderate, +++ = marked) after each item. Take periodic observations until the animal has returned to normal.

Time observed					
Number dead/number dosed					
1. Activity increase					
2. Activity decrease					
3. Ataxia					
4. Coma					
5. Convulsion(s)					
6. Corneal reflex loss					
7. Cyanosis					
8. Defecation, excessive					
9. Heart rate increase					
10. Heart rate decrease					
11. Irritability increase					
12. Irritability decrease					
13. Lacrimation increase					
14. Lacrimation decrease					
15. Muscle tone increase					
16. Muscle tone decrease					
17. Pain reflex loss					
18. Palpebral opening increase					
19. Palpebral opening decrease					
20. Piloerection					
21. Respiratory rate increase					
22. Respiratory rate decrease					
23. Righting reflex loss					
24. Salivation increase					
25. Startle response increase					
26. Startle response decrease					
27. Straub response					
28. Touch response increase					
29. Touch response decrease					
30. Tremors, twitches, jerks					
31. Urination increase					
32. Vasoconstriction increase					
33. Vasodilation increase					
34. Vocalization increase					

EXP. 10: OBSERVATION SHEET 2
(saline-treated)

Name________________________

Date________________________

Fill in the information below:

species__________ route__________ % sol./susp.__________
dose__________ drug__________ vehicle__________

Weigh the animal and record the weight, calculate and administer the appropriate dose, and record the volume administered and the time.

Weight of animal (kg) __________
Volume administered (cc) __________
Time dosed __________

Observe the animal (see Appendix D for definitions and observation techniques). Record the degree of response (0 = normal, + = slight, ++ = moderate, +++ = marked) after each item. Take periodic observations until the animal has returned to normal.

Time observed					
Number dead/number dosed					
1. Activity increase					
2. Activity decrease					
3. Ataxia					
4. Coma					
5. Convulsion(s)					
6. Corneal reflex loss					
7. Cyanosis					
8. Defecation, excessive					
9. Heart rate increase					
10. Heart rate decrease					
11. Irritability increase					
12. Irritability decrease					
13. Lacrimation increase					
14. Lacrimation decrease					
15. Muscle tone increase					
16. Muscle tone decrease					
17. Pain reflex loss					
18. Palpebral opening increase					
19. Palpebral opening decrease					
20. Piloerection					
21. Respiratory rate increase					
22. Respiratory rate decrease					
23. Righting reflex loss					
24. Salivation increase					
25. Startle response increase					
26. Startle response decrease					
27. Straub response					
28. Touch response increase					
29. Touch response decrease					
30. Tremors, twitches, jerks					
31. Urination increase					
32. Vasoconstriction increase					
33. Vasodilation increase					
34. Vocalization increase					

EXPERIMENT 11

Additional Procedures and Techniques in the Dog

EXPERIMENT 11A. Nail Trimming, Emptying Anal Sacs, Ear Cleaning, and Application of Eardrops in the Dog

INTRODUCTION

This experiment acquaints the student with nail trimming, emptying anal sacs, ear cleaning, and applying eardrops in dogs.

Many dogs and other animals never need their nails trimmed. But some animals need trimming for medical or cosmetic reasons or to prevent damage to the owner or the owner's property.

The anal sacs of dogs sometimes do not function normally and become irritating to the animal. Anal gland secretions must then be discharged manually.

The ear canals of dogs sometimes require cleaning and/or the application of eardrops.

Students work in pairs. One student performs the procedure while the lab partner restrains the animal, using mild to moderate restraint (see Exps. 8A, 8B). Then partners switch roles and repeat the procedure.

MATERIALS AND EQUIPMENT (per 2 students)

- dog
- flashlight, pencil
- nail clippers, dog
- silver nitrate applicator sticks (1 package per 20 students)
- gauze pads, 4 in. x 4 in. (1 package per 20 students)
- bandage, Conform stretch tape (1 roll per 20 students)
- 2 lab coats or jackets
- plastic disposable examination gloves, 2 pair
- cotton balls (1 package per 20 students)
- Vaseline (3 3/4 oz jar per 20 students)
- 10 swabs, cotton-tipped applicator
- 2 microscope slides (optional)
- 30 cc mineral oil, light, or miticide (Cerumite)
- beaker, 25 cc
- medicine dropper

PROCEDURE

Nail Trimming

Place the dog on the table. Have your lab partner hold the dog while you examine its nails. If the animal has a white nail, examine it carefully using the flashlight to transilluminate the nail. A blood vessel should be easily seen inside the white nail. The nail should be cut slightly beyond the end of this blood vessel, or it will bleed and cause the animal pain.

If the animal has all black or dark-colored nails, the end of the blood vessel must be estimated. Before making this estimate, examine a classmate's dog that has a white nail.

Your partner should restrain the dog with one hand gently around the dog's muzzle. If you suspect the animal might bite, place a gauze muzzle around its mouth (Exp. 8B).

Pick up the nail clippers and cut off the portion of nail below the end of the blood vessel (Fig. 11.1).

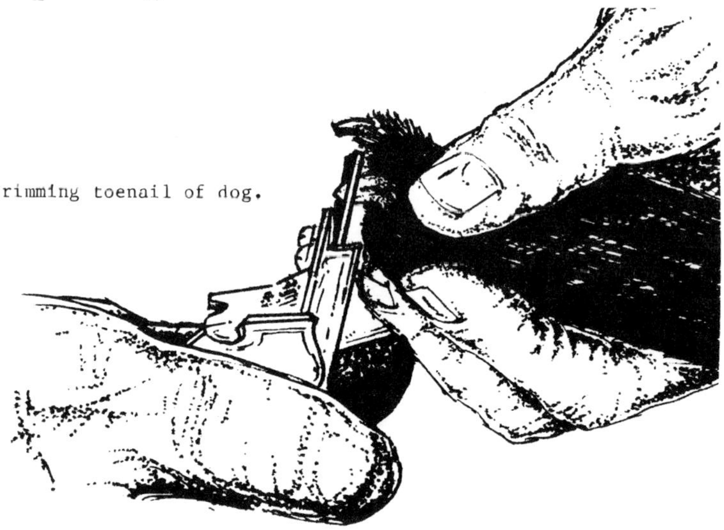

FIG. 11.1. Trimming toenail of dog.

If the nail is cut too short and bleeding occurs, hold the coated end of the silver nitrate stick against the bleeding toenail for several seconds. Usually this will cauterize the blood vessel. Occasionally a gauze pad and a Conform pressure bandage must be applied to stop the bleeding.

Trim the nails that appear long on two of the dog's feet and then switch positions so your partner can trim the remaining long nails.

Emptying Anal Sacs

Anal glands in dogs secrete an unpleasant smelling liquid that is stored in two anal sacs located on both sides of the anus. These normally empty into the sides of the anus. Occasionally the sacs do not empty properly and will cause the animal irritation and annoyance; then the anal sacs must be expressed manually.

A lab coat or jacket should be worn to protect your clothing, because this fluid has a strong and offensive smell.

Have your partner restrain the dog in a standing position. Usually an arm around its neck will be enough restraint. But if you suspect the dog will bite, use a gauze muzzle.

Put on plastic gloves. Apply Vaseline to your right gloved index finger. Pick up several cotton balls and hold them in the palm of your right hand with your lower two fingers. The cotton will absorb the expressed fluid.

Insert the index finger into the rectum about 1/2 in. by gently pressing it against the anus. Once in the rectum, rotate the finger so that the bottom of the finger is against one lateral wall. Then apply pressure against the rectum wall. This should cause fluid to squirt out of the sac into the cotton.

Then reverse roles with your partner and repeat the procedure on the other anal sac.

Ear Cleaning

Many animals have ear problems, with different causes. For example, the ear mite, Otodectes cynotis, causes inflammation and irritation in the ear canals of both cats and dogs. To treat an animal that has ear mites, the wax and debris in the ear canals and pockets must be cleaned out first and medication then applied.

Have your lab partner hold the dog's head firmly against his or her body.

To clean the outer ear, hold up the pinna (ear flap) in your left hand, pick up the cotton swab with your right hand, and insert the swab gently into the outer ear. Swab out any ceruminous material present.

The ear canal is cleaned next. The ear canal is a funnel-shaped tube that descends and terminates at the tympanic membrane (Fig. 11.2).

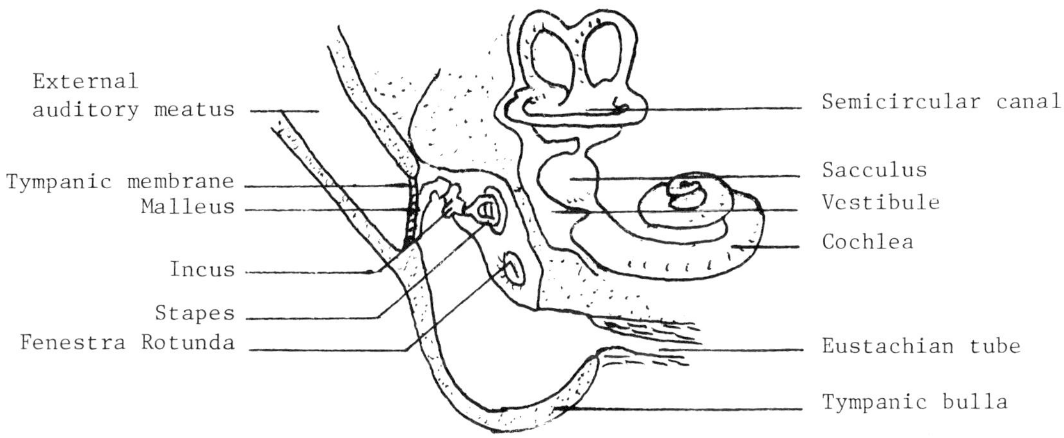

FIG. 11.2. Ear canal terminating at tympanic membrane.

Care must be taken not to damage the canal or puncture the tympanic membrane. Place the swab very gently in the ear canal and swab the area. Withdraw the swab and proceed with a clean swab if needed. Then trade places so your partner can clean the other ear canal.

The extracted ceruminous material can be checked for mites by rolling the swab on a microscope slide. Several drops of microscope oil should be added before examining the material. Use low power.

Application of Eardrops

Either mineral oil or a miticide is used. Restrain the dog as for ear cleaning. If using mineral oil, pour about 15 cc of the oil into the 25 cc beaker. Fill the medicine dropper with oil (to approximately 1/4 in. below the rubber bulb) and discharge it into the external auditory meatus.

Massage the external lower area of the ear gently to distribute the oil thoroughly into the lower portion of the canal.

If using a miticide, follow the manufacturer's instructions.

Reverse roles with your partner and clean and medicate the other ear.

EXPERIMENT 11B. Oral Administration of Acepromazine Maleate in the Dog

INTRODUCTION

This experiment acquaints the student with administering tablets orally to dogs and with the effects of acepromazine maleate (Atrovet, Notensil) in dogs.

Acepromazine is primarily used as a preanesthetic agent or tranquilizer in dogs.

Students work in pairs. One student monitors vital signs while the lab partner restrains the animal, using mild restraint (see Exp. 8A). Then partners reverse roles and repeat the procedures. For administration of tablets, the dose is divided in half, and each student administers one portion of the dose.

MATERIALS AND EQUIPMENT (per 2 students)

- human scale (per 10 students)
- dog
- lubricating agent (soap, Vaseline, etc.) (1 jar per 20 students)
- thermometer, rectal
- 4 cotton balls or tissues
- thermometer holder with alcohol (optional)
- 50 cc bottle of alcohol, isopropyl, 70% (per 20 students) (if thermometer holder is not used)
- acepromazine maleate (Atrovet or Notensil), 25 mg tablets (2-6 tablets per 2 students, depending on weights of dogs)

PROCEDURE

Weighing

Check your scale for accuracy and adjust if necessary (see Exp. 1). Weigh the dog (see Exp. 8B) and record the weight on the observation sheet.

Vital Signs

Determine and record on the student report sheet the dog's temperature, pulse, and respiration (see Exp. 8A).

Dose

The dose is 1 mg/lb. On your student report (numbers 2 and 3), calculate the number and/or portion of tablet(s) to give according to the dog's weight. Have your instructor check your calculations.

Administration of Tablets

Pick up the tablet(s) in your right hand and hold between index finger and thumb. Place your left hand over the dog's muzzle and with your left fingers and thumb, press the dog's upper lip against its teeth. Apply pressure and tilt its head backwards. With your middle right finger, press downward on its lower jaw (Fig. 11.3).

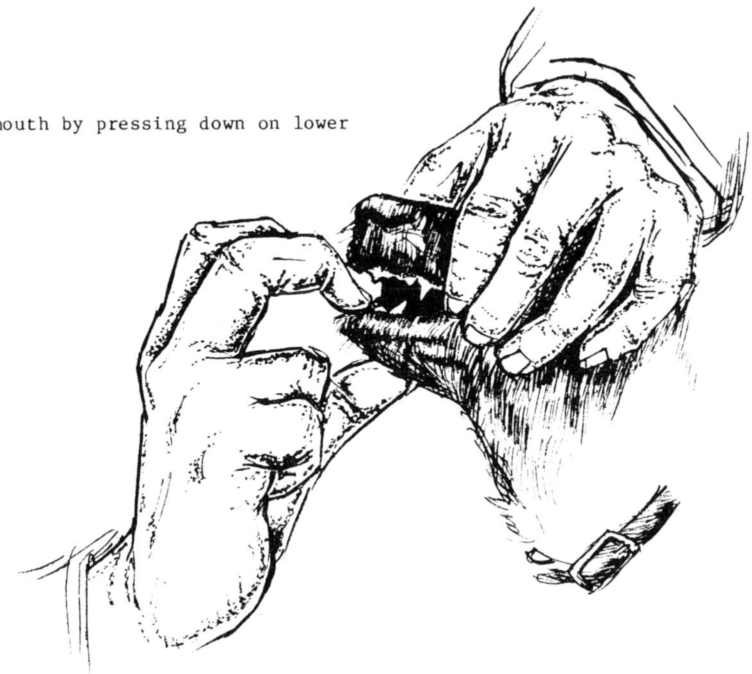

FIG. 11.3. Opening dog's mouth by pressing down on lower jaw with middle finger.

When the mouth is wide open, quickly push the tablet over the hump in the back (base) of the tongue. Close the mouth and with the dog's head still elevated, stroke the throat gently (Fig. 11.4). The animal usually swallows spontaneously.

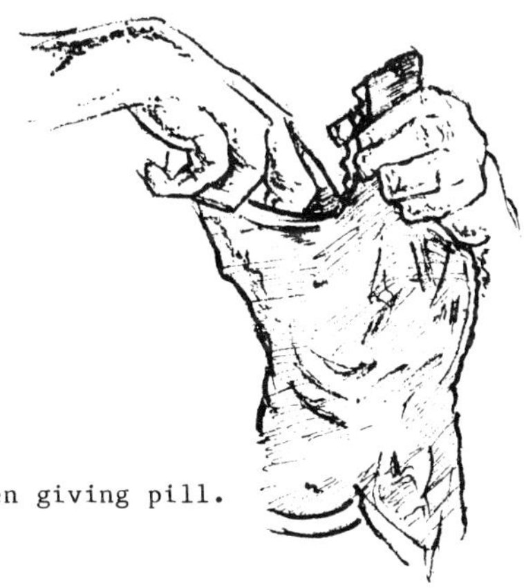

FIG. 11.4. Holding dog's head when giving pill.

Then your partner administers a second tablet. Take turns administering any remaining tablets in the dose.

Observe the animal and record your observations on the observation sheet. Take the rectal temperature, respiration, and pulse 1 and 2 hr postadministration of acepromazine and record the values on the student report.

EXP. 11B: STUDENT REPORT

Name______________________

Date______________________

1. Determine and record the following in the untreated dog:

   a. Temperature______________

   b. Respiration______________

   c. Pulse______________

2. Calculate how many milligrams to give your dog (show your work).

3. Calculate the number of tablets or the portion of tablet to give your dog (show your work).

4. a. After administering acepromazine maleate, determine and record the following:

	1 hr postadministration	2 hr postadministration
Temperature	______________	______________
Respiration	______________	______________
Pulse	______________	______________

   b. Explain your results.

5. a. For a 0.25 mg/lb dose of acepromazine maleate, calculate how many milligrams to give your dog (show your work).

   b. According to your answer in 5a, how many tablets or what portion of a tablet should you give your dog?

6. a. For a 1.35 mg/kg dose, calculate how many milligrams your dog should receive (show your work). (Remember, you must change either kilograms to pounds or pounds to kilograms.)

   b. According to your answer in 6a, how many tablets or what portion of a tablet should you give your dog? (Show your work.)

EXP. 11B: OBSERVATION SHEET

Name______________________________

Date______________________________

Fill in the information below:

species__________ route__________ % sol./susp.__________
dose__________ drug__________ vehicle__________

Weigh the animal and record the weight, calculate and administer the appropriate dose, and record the volume administered and the time.

Weight of animal (lb) ____________
No. of tablets administered ____________
Time dosed ____________

Observe the animal (see Appendix D for definitions and observation techniques). Record the degree of response (0 = normal, + = slight, ++ = moderate, +++ = marked) after each item. Take periodic observations until the animal has returned to normal.

Time observed					
Number dead/number dosed					
1. Activity increase					
2. Activity decrease					
3. Ataxia					
4. Coma					
5. Convulsion(s)					
6. Corneal reflex loss					
7. Cyanosis					
8. Defecation, excessive					
9. Heart rate increase					
10. Heart rate decrease					
11. Irritability increase					
12. Irritability decrease					
13. Lacrimation increase					
14. Lacrimation decrease					
15. Muscle tone increase					
16. Muscle tone decrease					
17. Pain reflex loss					
18. Palpebral opening increase					
19. Palpebral opening decrease					
20. Piloerection					
21. Respiratory rate increase					
22. Respiratory rate decrease					
23. Righting reflex loss					
24. Salivation increase					
25. Startle response increase					
26. Startle response decrease					
27. Straub response					
28. Touch response increase					
29. Touch response decrease					
30. Tremors, twitches, jerks					
31. Urination increase					
32. Vasoconstriction increase					
33. Vasodilation increase					
34. Vocalization increase					

EXPERIMENT 12

The Effects of Ovariectomy and Estradiol Cypionate on the Uterus of the Rat

EXPERIMENT 12A. Determination of Estrous Cycle Phases, Ovariectomy, and Administration of Estradiol Cypionate in the Rat

INTRODUCTION

This experiment runs for a 15-day period. On the first day students divide the animals into the following four groups:

group 1--normal, untreated controls
group 2--normal, ECP-treated animals
group 3--ovariectomized, untreated animals
group 4--ovariectomized, ECP-treated animals

Each student takes vaginal cell samples from two animals, makes and stains vaginal smears, and determines the estrous cycle phases using the microscope.

With students working in pairs, each student administers anesthesia and performs an ovariectomy on one animal in group 3 or 4.

On day 8, working in pairs, students remove the stitches. Then estradiol cypionate (ECP), a female hormone, is administered daily by each student for 7 days, SC, to one animal in group 2 or 4. Estrous cycle phases are also determined and recorded for all animals during this 7 day period.

Experiment 12B completes the experiment on the 15th day by determining the effects of the ovariectomies and EPC administration on the uteri.

MATERIALS AND EQUIPMENT

Vaginal Smears (per student)

sterile saline, 15 cc
2 female rats, each approximately 200 g
beaker, 20 cc
3 cotton-tipped applicator swabs
2 microscope slides
Diff-Quick Stain Set (per 10 students)
3 Coplin jars (per 10 students)
beaker with 500 cc distilled or deionized water (per 10 students)
microscope
marker, indelible (or leather punch) (per 10 students)

Ovariectomy (per 2 students)

anhydrous ether, anesthesia grade (1 pt per 20 students)
hood (optional) (per 8 students)
ether jar (per 4 students)
cotton (1 roll per 20 students)
beaker, 20 cc
clippers, surgical, small animal (per 4 students)
4 paper towels
2 pairs of surgical gloves, sterile
scalpel with blade, size 20
forceps, rat-tooth, large
surgical scissors, blunt nose, large
2 hemostats
4 sterile gauze pads, 2 in. x 2 in.
suture needle
surgical suture, cotton or nylon, size 00
needle holder
empty cage, clean

Suture Removal on 8th Day (per 4 students)

scissors, stitch (suture removal)
forceps, rat-tooth, large

ECP Injection on 8th Day (per student)

estradiol cypionate (ECP), 2 mg/ml (1 vial per 20 students)
olive oil, sterile (20 cc per 20 students)
syringe, 1 cc
needle, 22 gauge, 1/2 in.
10 cc graduated cylinder
10 cc bottle

PROCEDURE

Experimental Design

Divide the animals evenly into the following groups and mark the cages accordingly:

group 1--normal, untreated controls
group 2--normal, ECP-treated animals
group 3--ovariectomized, untreated animals
group 4--ovariectomized, ECP-treated animals

It is best to use individual cages if they are available. But if they are not available, group cages can be used, preferably five rats per cage; but the animals must then be marked so that each animal can be identified. An indelible, quick-drying marker can be used. Animal 1 can be marked on the head, 2 on the middle of the back, 3 on the back just above the tail, 4 on the head and tail, and 5 can be left unmarked. Or holes can be punched in their ears with a leather punch, using a similar-type code.

Each student selects two rats, one each from groups 1 and 4 or groups 2 and 3, to use for his or her portion of the experiment. Each student works individually and makes a vaginal smear from each of the two rats.

Vaginal Smears

The estrous cycle lasts about 5 days in the rat. Stages of the estrous cycle are identified by microscopic examination of vaginal cells (vaginal smears). It is important to know specific stages in order to mate animals for timed deliveries.

The vaginal smear procedure is relatively easy. Pour about 15 cc sterile saline into the 20 cc beaker. Take a cotton swab in your right hand and moisten it in the saline. Pick up a rat midtail with your left hand and place it on the lab table top, still holding the tail. Insert the swab gently into the vagina and rotate it slightly (Fig. 12.1). This sampling procedure does not seem to bother the rat.

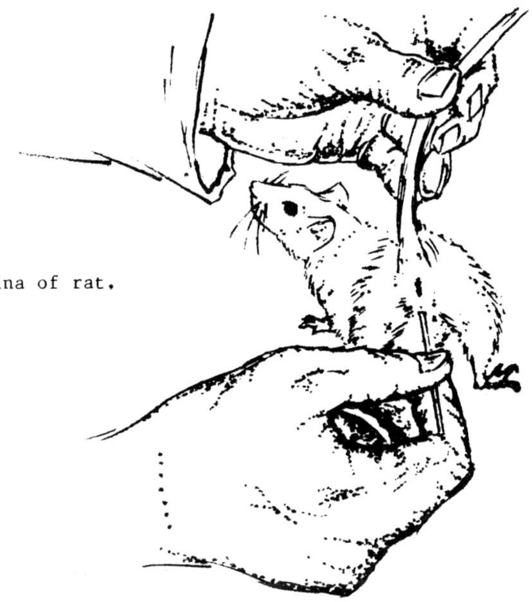

FIG. 12.1. Inserting cotton swab into vagina of rat.

Remove the swab and replace the rat in the cage.

Roll the swab gently along a clean microscope slide. Allow the slide to air dry.

Repeat this procedure on your second rat.

Staining Cells

Remove the three bottles of solutions from the Diff-Quik box and line them up on the workbench near a sink with the Fixative first, Solution 1 next, and Solution 2 last. Place the three Coplin jars in front of each bottle and fill the jars with the appropriate solutions. Replace the bottles behind the jars.

Fill a beaker with 50 cc distilled or deionized water and place it near the sink.

Holding the air-dried smear by one end, lower it into the Fixative solution 5 times for 1 sec per dip. Allow the excess solution to drip into the Fixative jar. Dip the slide 5 times in Solution 1 for 1 sec per dip; allow the excess to drip into the jar. Dip the slide 5 times into Solution 2, 1 sec per dip; allow the excess to drip.

Then hold the slide over the sink and gently pour the distilled water against the top of the slide to rinse off the extra stain. Allow the water to flow across the smear. (The water should not strike the smear directly.)

Let the rinsed smear air dry.

Replace the tops on the Coplin jars.

Estrous Cycle Phases and Vaginal Cell Types

The following are the estrous cycle phases and the cells found in each phase:

1. Estrus: period of sexual receptivity lasting about 12 hr; indicated by large cornified (degenerative cells that lose nuclei) epithelial cells, some nucleated (Fig. 12.2).

2. Metestrus: period shortly after ovulation lasting about 21 hr; indicated by large cornified epithelial cells mixed with polymorphonuclear leukocytes (about 1/3 the size of epithelial cells).

3. Diestrus: period of inactivity lasting about 60-70 hr; indicated by mostly polymorphonuclear leukocytes and a few nucleated epithelial cells.

4. Proestrus: preparatory period for next estrous phase lasting 12 hr; indicated by round, nucleated epithelial cells and a few leukocytes.

Look at the dried, stained smears under the microscope, using low power, and determine the estrous phase of each rat. Record the phases on your student report sheet.

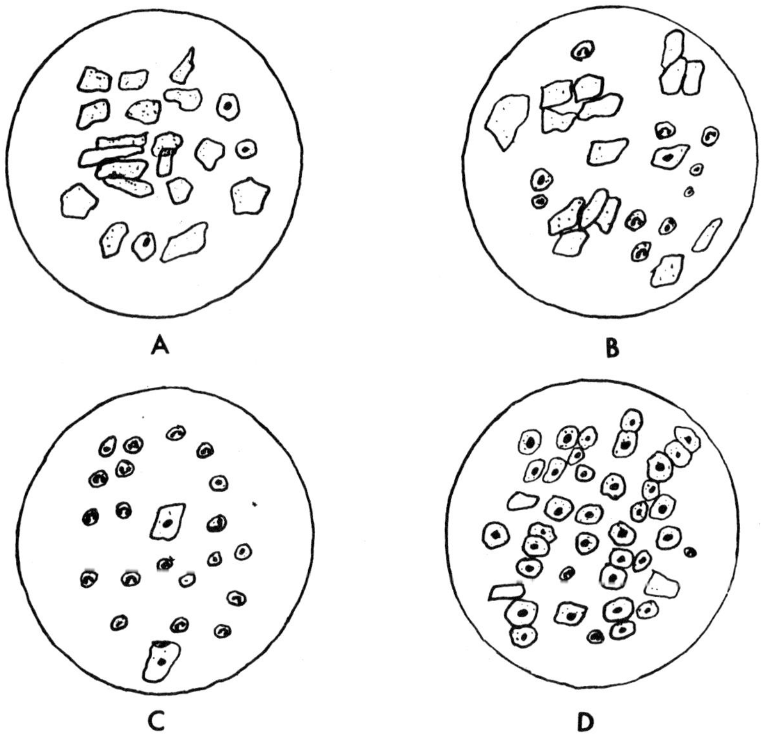

FIG. 12.2. Estrous cycle phases and vaginal cell types: A. Estrus: large cornified epithelial cells; a few round, nucleated epithelial cells. B. Metestrus: cornified epithelial cells; many leukocytes. C. Diestrus: mostly leukocytes; a few nucleated epithelial cells. D. Proestrus: nucleated, circular epithelial cells, a few leukocytes (about 1/3 size of epithelial cells).

Anesthesia Administration

Ether will be used as the anesthesia. Ether is an explosive, flammable substance, and ether gas is heavier than air and will flow along the floor. Before using ether, all doors should be closed and "No Smoking--Explosive" signs posted at all entrances to the lab. Bunsen burners should not be used, or any electrical equipment that might cause a spark. (Animal clippers are especially grounded to prevent sparks.)

Work in pairs, one student acting as anesthesiologist and the other as surgeon for one ovariectomy and then reversing roles for the second ovariectomy.

Place about 1 in. of cotton in the bottom of the ether jar. If a hood is available, the initial anesthesia administration is conducted in the hood. Pour enough ether to wet the bottom of the jar. Place about 1/3 in. of cotton in the bottom of a 20 cc beaker, making a nose cone, and pour in enough ether to wet the bottom. Position the beaker upside down (so the ether doesn't evaporate) near the surgical area.

Place the rat quickly in the ether jar and secure the lid (quickness is the trick). Observe the rat carefully and remove it from the jar when it has lost its righting reflex.

Your partner holds the animal while you shave the midline area of the abdomen with surgical clippers (try to do this over a wastepaper basket). Place the animal on a clean paper towel and put the nose cone over its nose. The cotton should not touch the nose. The anesthesiologist must observe the patient at all times. Do not become distracted from your patient by the surgery. Anesthesia is regulated manually by moving the nose cone closer to the nose for more anesthesia or farther from the nose for less anesthesia. (If you need more anesthesia and have the nose cone as close to the nose as possible, you might need to add more ether to the nose cone.)

Ovariectomy

Put on the surgical gloves and, using the scalpel, make a ventral midline incision through the skin, starting midway in the abdomen and running caudally about 1 1/2 in. (Fig. 12.3).

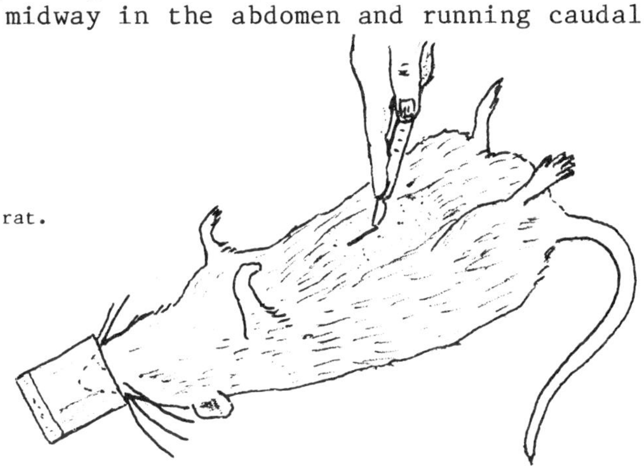

FIG. 12.3. Midline incision for ovariectomy in rat.

Use the forceps to pick up the abdominal rectus muscle, and cut through the muscle with the surgical scissors. Make sure you do not cut into the intestines.

The ovaries are embedded in fat, lying just caudally to the kidneys. (If you cannot locate the ovaries, locate the uterus near the bladder and follow it cranially to the ovaries.)

Gently free the fat from the ovary using the scissors. Clamp the fallopian tube with a hemostat just below the ovary. Using the suture, place a ligature around the tube below the hemostat and tie firmly in a square knot. Remove the ovary by cutting through the fallopian tube with the scalpel between the hemostat and the ovary. Release the hemostat and replace the fallopian tube in the abdominal cavity. Repeat this procedure with the second ovary.

Thread your needle and place it in your needle holder. Starting cranially, with the forceps in your left hand, pick up the peritoneum (it will look shiny) and muscle layers together at the incision site on the rat's left side. With your right hand holding the needle holder, insert the needle through the muscle and the peritoneum of the left side of the incision. Release the needle from the needle holder and then pull the needle through with the needle holder. Repeat with the other side of the incision, inserting the needle first through the peritoneum and then through the muscle layers. Pull the two sides of the incision together with the suture and tie in a square knot. (Some surgeons add another half-knot for security.) Cut the suture fairly close to the knot (about 3 mm). Place additional sutures approximately every 1/4 in.

After closing the peritoneum and muscle layers, close the skin in the same manner but tie double knots and leave slightly longer tails on the knots (approximately 1/4 in.).

Place the patient in a clean cage and allow to recover for 1 wk.

Reverse roles with your lab partner and perform an ovariectomy on a second rat, with your lab partner acting as surgeon.

Suture Removal

On the 8th day, remove all sutures. It is easier for inexperienced students to work in pairs to do this. Have your lab partner restrain the rat so that the abdominal area is exposed. Slide the hooked portion of the stitch scissors under the suture and cut it. Then, using forceps, pull the knotted end of the suture so the suture comes out. Repeat this procedure.

ECP Administration

On the 8th day postoperatively and daily thereafter for 7 days, treat your animal in group 2 or 4 daily SC with ECP (each rat in groups 2 and 4 receives 0.015 mg/kg ECP by SC injection).

On the student report (number 2), calculate the milligrams of ECP to give a 200 g rat (answer: 0.003 mg).

The concentration of ECP in the vial is 2 mg/cc. On the student report (number 3), calculate the volume to give a 200 g rat (answer: 0.015 cc).

This volume is too small to administer. Thus the concentration must be diluted. Make a 1:100 dilution of ECP to olive oil. Use the 1 cc syringe to withdraw and measure 0.1 cc ECP. Place in the 10 cc graduated cylinder and *q.s.* to 10 cc with olive oil. Pour this dilution carefully into the 10 cc bottle. On your student report (number 4), calculate the volume of this dilution to give a 200 g rat (answer: 0.15 cc). Have your instructor check your calculations.

If your animals weigh about 200 g, this one dilution should last about 5 days for 20 students. Several students should make this initial dilution, and if more is needed, several others should make the next dilution when needed.

Administer the diluted ECP solution SC daily to your group 2 or 4 animal, using a 1 cc syringe with a 20 gauge needle (a large needle is needed because the oil is viscous).

Daily Vaginal Smears

On the 8th day postoperatively and daily thereafter for 7 days, determine and record the estrous cycle phases of your two rats.

EXPERIMENT 12B. Euthanasia, Dissection of the Uterus, and Determination of the Uterus Weight of the Rat

INTRODUCTION

This experiment acquaints the student with euthanasia, dissection of the uterus, and using a balance to determine organ weights. The effects of ovariectomies and ECP on the uterus are determined.

MATERIALS AND EQUIPMENT (per 2 students)

- anhydrous ether (1 pt per 20 students)
- hood (optional) (per 8 students)
- ether jar (per 4 students)
- beaker, 20 cc
- cotton (1 roll per 20 students)
- surgical scissors, blunt nose
- surgical scissors, small
- rat-tooth forceps, large
- balance (Mettler, 1.000 g or equivalent) (per 4 students)
- 4 pieces weighing paper or 1 weighing dish

PROCEDURE

Euthanasia

On the 15th day postoperatively, after determining their estrous cycle phases, sacrifice both rats by an overdose of ether. Administer ether as was done in Exp. 12A, but leave the animal in the jar until it is no longer breathing.

While you are waiting for the ether to take effect, tare the weighing paper or dish (weigh it so you can subtract its weight from the total weight of the paper or dish and the organ).

Dissection of Uterus

Remove the euthanized rat from the ether jar and place an ether cone over its nose to prevent resuscitation. Using the larger surgical scissors and forceps, make a midline incision through the skin and muscle layers. Locate the uterus as in Exp. 12A and remove it by cutting through the uterus as close to the vagina as possible.

Determination of Uterus Weight

Place the uterus on the weighing paper or dish and, using the smaller scissors and forceps, trim as much fat from the organ as possible. Then weigh the tared paper or dish containing the uterus. Subtract the weight of the paper or dish, and record the weight of the uterus on your student report sheet.

Repeat this procedure with your second rat. Then record both uterus weights on the class blackboard, under the appropriate group. Record all the weights of each group on your student report sheet.

Complete your student report sheet.

EXPS. 12A, 12B: STUDENT REPORT

Name______________________________

Date______________________________

1. Record the group number, animal number (if you didn't use individual cages), and estrous cycle phases of your animals:

	Group __________	Group __________
	Rat no. __________	Rat no. __________
	Estrous phase	Estrous phase
Day 1 (preoperative)	__________	__________
Day 8	__________	__________
Day 9	__________	__________
Day 10	__________	__________
Day 11	__________	__________
Day 12	__________	__________
Day 13	__________	__________
Day 14	__________	__________
Day 15	__________	__________

2. The dose of ECP to give is 0.015 mg/kg. Calculate the milligrams of ECP to give a 200 g rat (show your work).

3. The concentration of ECP in the vial is 2 mg/cc. Calculate the volume of ECP to give a 200 g rat (show your work).

4. Calculate the volume of a 1:100 dilution of ECP to give a 200 g rat (show your work).

5. Record the weight of the uteri of your rats in the appropriate column and circle them. Then copy all the other uteri weights from the class blackboard (add more lines if there are more than 20 in your class), and compute the average weight of each group.

Group 1 Untreated Controls	Group 2 ECP-treated Controls	Group 3 Ovariectomized Untreated Animals	Group 4 Ovariectomized ECP-treated Animals
Averages			

6. Explain the results in number 5.

EXPERIMENT 13

The Effects of Castration and Testosterone on the Seminal Vesicles of the Rat

INTRODUCTION

This experiment acquaints the student with performing nonsterile castration in rats and the effects of testosterone, the male hormone, in maintaining the size of the seminal vesicles.

MATERIALS AND EQUIPMENT

Castration (per 2 students)

- 2 male rats, each approximately 200 g
- anhydrous ether, anesthesia grade (1 pt per 20 students)
- hood (optional) (per 8 students)
- ether jar (per 4 students)
- cotton (1 roll per 20 students)
- beaker, 20 cc
- 2 pair surgical gloves, sterile
- scalpel with blade, size 20
- 3 hemostats, curved
- surgical suture, nylon, size 00 (optional)
- needles, suture, size C-40 (optional)
- needle holder (optional)
- surgical scissors, blunt nose
- rat-tooth forceps

Injections, SC (per student)

- 3 male rats, each approximately 250 g
- 2 cc testosterone suspension, 100 mg/ml
- sterile bottle with 50 cc sterile saline (per 20 students)
- 15 needles, 22 gauge, 1/2 in.
- 2 syringes, 1 cc

Dissection (per student)

- anhydrous ether (1 pt per 20 students)
- hood (optional) (per 8 students)
- ether jar (per 4 students)
- beaker, 20 cc
- cotton (1 roll per 20 students)
- surgical scissors, blunt nose
- surgical scissors, small
- rat-tooth forceps, large
- rat-tooth forceps, small
- 2 weighing papers or dishes
- balance (Mettler, 1.000 g or equivalent) (per 4 students)

PROCEDURE

Experimental Design

Divide the animals evenly into the following groups and mark the cages accordingly:

group 1--uncastrated, saline-injected animals
group 2--castrated, saline-injected animals
group 3--castrated, testosterone-injected animals

Use individual cages if they are available. If not available, use group cages and mark the individual animals (see Exp. 12A).

Each student castrates one animal from group 2 or 3 and then administers testosterone or saline for 14 days. In addition, the instructor assigns the remaining animals in group 1 to one-half of the students for injection of saline. The remaining one-half of the students are then assigned these animals at the termination of the experiment for dissection, weighing, and recording the weights of the seminal vesicles.

Castration

Work in pairs, one student acts as anesthesiologist and the other as surgeon for one castration; reverse roles for the second castration.

Your partner anesthetizes the rat (see Exp. 12A) and places it in a dorsoventral position while continuing to administer anesthesia.

Put on surgical gloves and locate both testes in the scrotum. If the testes are not visible they may be retracted into the abdominal cavity; slight pressure on the abdomen will force them into the scrotum. With your left fingers still applying slight pressure on the abdomen, pick up your scalpel in your right hand and make one ventral, midline incision through the scrotum.

Retract the skin to make the testes visible. If the testes are not completely visible, you have not cut through the tunica vaginalis that covers the testes. Make another incision through this tissue if necessary.

Withdraw one testis through the incision. The anterior end of the testis is attached to the rat's body by the spermatic artery and vein; the caudal end is attached by the vas deferens (Fig. 13.1).

Hold the vein, artery and vas deferens with the rat-tooth forceps (not with the teeth but with the inside of the prongs) and, using the small scissors, free as much fat as possible.

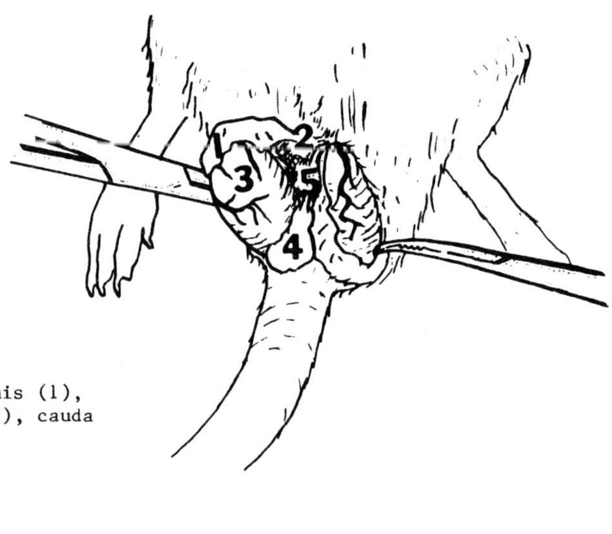

FIG. 13.1. Dissection of rat testes: caput epididymis (1), right spermatic artery and vein (2), right testis (3), cauda epididymis (4), right vas deferens (5).

Then clamp them together with a curved hemostat. Using the suture, place a single ligature around the vessels anteriorly to the hemostat (Fig. 13.2). Tie the square knot firmly and add a second square knot as a safety precaution. Cut the vessels caudally to the hemostat and remove the testis. Cut the excess suture close to the knot.

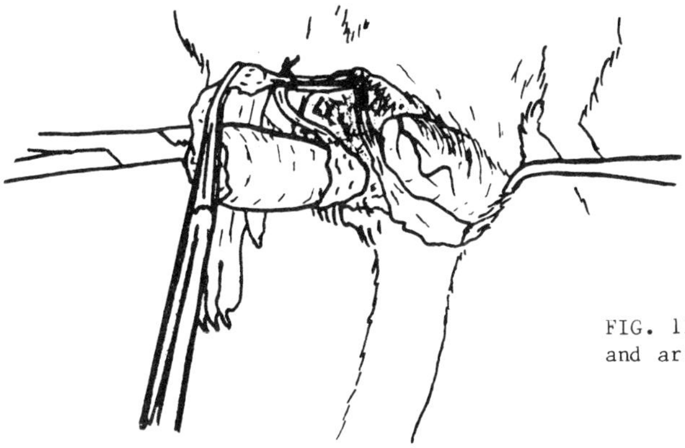

FIG. 13.2. Placing tie around vas deferens, spermatic vein, and artery.

Repeat this procedure on the second testis.

Reverse roles with your partner and castrate a second rat.

The incision in the skin may be closed by suturing (see Exp. 12A) or left open to drain. If sutured, the stitches should be removed on day 7.

Testosterone and Saline Injection, SC

Give daily injections SC of 0.1 cc testosterone or saline for 14 days. Use one syringe for the testosterone-treated animal and the other for the control animal. Replace the needles after several injections because they become dull.

Dissection of Seminal Vesicles

On day 15 tare a piece of weighing paper (see Exp. 12B), and place it where you will be dissecting your animal.

Euthanize the rats (Exp. 12B). Place your animal in a dorsoventral position and make a midline incision from the umbilicus to the scrotum, exposing the intestines and bladder. On either side of the bladder lies a pale structure about 1 in. long and 1/4 in. wide. This structure consists of two parts, the seminal vesicle and the smaller coagulatory gland. They both narrow down to thin stalks where they connect to the vas deferens. The two parts need not be separated. Dissect caudally as far as possible, remove at the stalk, and place on the tared weighing paper (Fig. 13.3).

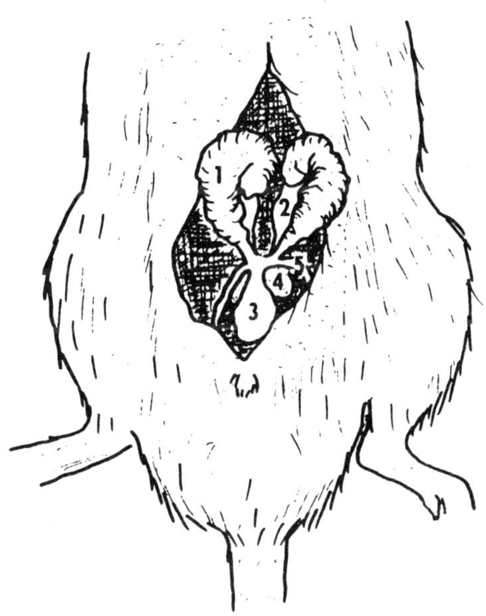

FIG. 13.3. Location of seminal vesicle and coagulating gland in rat: seminal vesicle (1), coagulating gland (2), bladder (3), prostate gland (4), vas deferens (5).

Repeat with the other seminal vesicle and coagulatory gland. Weigh both structures together. Subtract the weight of the paper from the weight of the organs plus paper to determine the weight of the organs.

Results

Record the organ weights of each animal on the class blackboard, according to their groups. Compute the average weight for each group. Record the class averages on your student report sheet and complete the report sheet, explaining the results.

EXP. 13: STUDENT REPORT

Name______________________________

Date______________________________

1. Record the results:

Group	Average wt of seminal vesicles and coagulatory glands (mg)
1 (uncastrated, saline-injected)	
2 (castrated, saline-injected)	
3 (castrated, testosterone-injected)	

2. Explain the results.

EXPERIMENT 14

Common Procedures and Techniques in the Bird

INTRODUCTION

This experiment acquaints the student with restraining and collecting blood from birds, administering medication intramuscularly (IM) and orally in birds, and clipping nails and trimming wings of birds.

Students work in pairs and take turns performing the following procedures: restraining, wing trimming, weighing, collecting blood using two different techniques, IM injection, oral administration, and nail clipping.

MATERIALS AND EQUIPMENT (per 2 students)

Restraint
- 1 pigeon (or another species of bird)
- 1 towel or washcloth (optional)

Wing Trimming
- scissors, 5 1/2 in., straight, blunt

Weighing
- bird scale and balance weights (per 4 students)

Blood Collection, Toenail
- 4 cotton balls
- 50 cc bottle of alcohol, isopropyl, 70%
- nail clippers, animal
- 2 microhematocrit tubes, heparinized
- 2 gauze squares, 2 in. x 2 in.
- sealing clay (Seal-Ease Tube Sealer and Holder) (per 4 students)
- 4 microscope slides
- silver nitrate applicator sticks (1 package per 20 students)

Blood Collection, Brachial Vein
- 2 needles, 23 gauge, 1/2 in.
- 2 syringes, 1 cc
- 2 centrifuge tubes, tapered, 13 x 100 mm (or equivalent)
- centrifuge rack or 100 cc beaker (per 4 students)

IM Injection
- 2 needles, 25 gauge, 1/4 in.
- 2 syringes, 1 cc
- beaker with 10 cc sterile saline
- 4 cotton balls
- 50 cc bottle of alcohol, isopropyl, 70%

Oral Administration

10 cc beaker of tap water
feeding needle, 18 gauge, 3 in.
syringe, 1 cc
forceps, thumb, 5 in.

Nail Clipping

nail clippers, animal
emery board
silver nitrate applicator sticks (1 package per 20 students)

PROCEDURE

Restraint

Physical restraint of birds is necessary to perform routine health care or treatment. Proper handling ensures safety for the bird as well as for the handler.

Before restraining any bird, make sure all the windows and doors are closed in case the bird escapes from restraint. Also turn off all fans or hoods.

A towel or washcloth may be used to cover the bird's head and body before it is picked up. The advantages of using a towel are that it minimizes panic in the bird and helps to keep the handler's hands (at 98.6°F) from overheating the bird, especially a small bird. Some birds, however, are easier to catch without a towel. After catching the bird, place your right hand over its back and trap the wings next to the body.

To hold the pigeon so you won't be scratched, transfer it to your left hand. Your left hand should be under its stomach with your little finger closest to its head (Fig. 14.1). Catch its legs between your index and second fingers and have your thumb over its back.

FIG. 14.1. Restraining bird.

Wing Trimming

Wing trimming is painless if performed properly. Birds that are allowed out of their cage should have their wings clipped for their own protection. When a bird is frightened its first reaction is to take flight and it may fly into a window and hurt itself or escape through an open window or door.

There are two methods of trimming wings. One method involves cutting the outermost eight or nine primary feathers on one wing only so the bird is asymmetrical and cannot fly. The second method is to trim both wings. The advantage of the second method is that the wings appear normal when folded. To trim both wings, have your lab partner restrain the bird. Extend one wing with your left hand and pick up the scissors in your right hand. Leave the

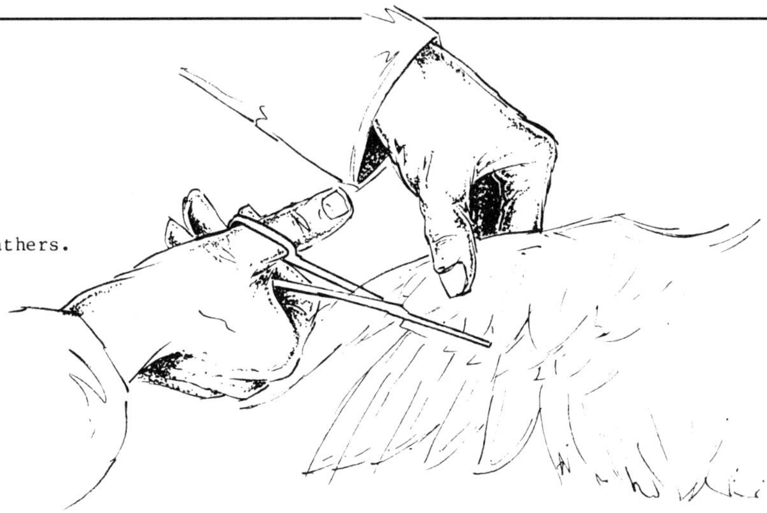

FIG. 14.2. Cutting primary feathers.

first three primary feathers intact. Cut the next eight primary feathers just beyond the tip of the covert feathers (Fig. 14.2). After trimming one wing, change roles with your lab partner and have your lab partner trim the feathers of the second wing.

Weighing

Check the scale for accuracy and adjust if necessary (See Exp. 1).

Weigh the bird by placing it in the weighing pan. Have your lab partner put the lid on as soon as you place the bird in the container.

Weigh the bird and record the weight on the student report sheet. Remove the pan from the scale and restrain the bird again.

Blood Collection

Toenail technique. If only several drops of blood are required, clipping the toenail is the least stressful method. Have your lab partner restrain the bird so that its claws are facing the ceiling (dorsal recumbency). Carefully clean the nail to be clipped with a cotton ball saturated in alcohol so that all dirt or droppings are removed. This is necessary because blood contaminated with urea may elevate uric acid determinations. Make sure the nail is completely dry of alcohol before proceeding to the next step. Pick up the nail clippers in your right hand and clip the nail just above the quick (between the nail base and the end of the blood vessel) so that the nail will bleed freely (Fig. 14.3). Don't hold the toe tightly because this will prevent bleeding.

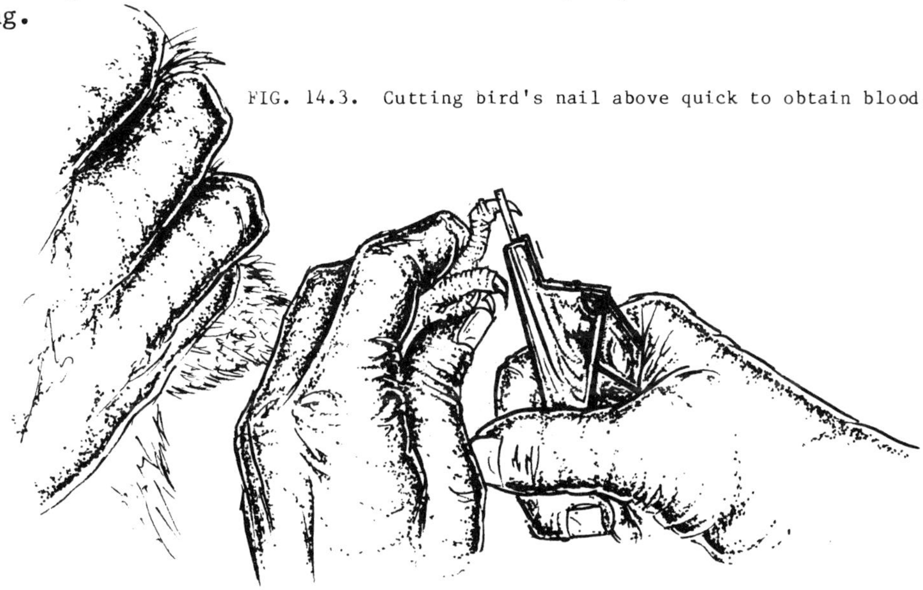

FIG. 14.3. Cutting bird's nail above quick to obtain blood.

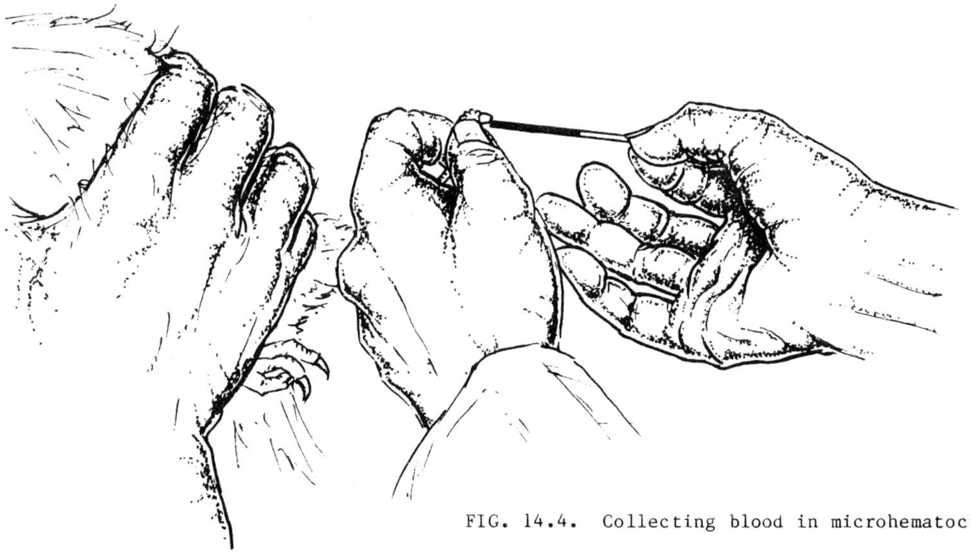

FIG. 14.4. Collecting blood in microhematocrit tube.

Have the sealing clay ready to hold the microhematocrit tube. Pick up the microhematocrit tube and hold it near one end between your right thumb and forefinger. If a Pre-cal tube is used, hold it nearest the calibration mark. Slightly elevate the far end of the tube (approximately 15°) and touch it to the drop of blood (Fig. 14.4). The blood should flow freely into the tube. Allow an uncalibrated tube to fill to approximately 1/8 in. from the end. Allow a Pre-cal tube to fill to the calibration mark. To stop the flow of blood into the tube, raise the end you are holding to make the tube horizontal. If you fill the tube too far, remove the excess by touching the tip that was filled gently to an absorbent-type paper or gauze.

Clean off the end of the blood-filled tube with a piece of gauze. Push the end that you were holding into the clay so that the tube stands upright (Fig. 14.5). Set this aside if you plan to do Exp. 15 within 90 min, or else discard (blood should be kept no longer than 90 min for these tests).

FIG. 14.5. Sealing microhematocrit tube with clay.

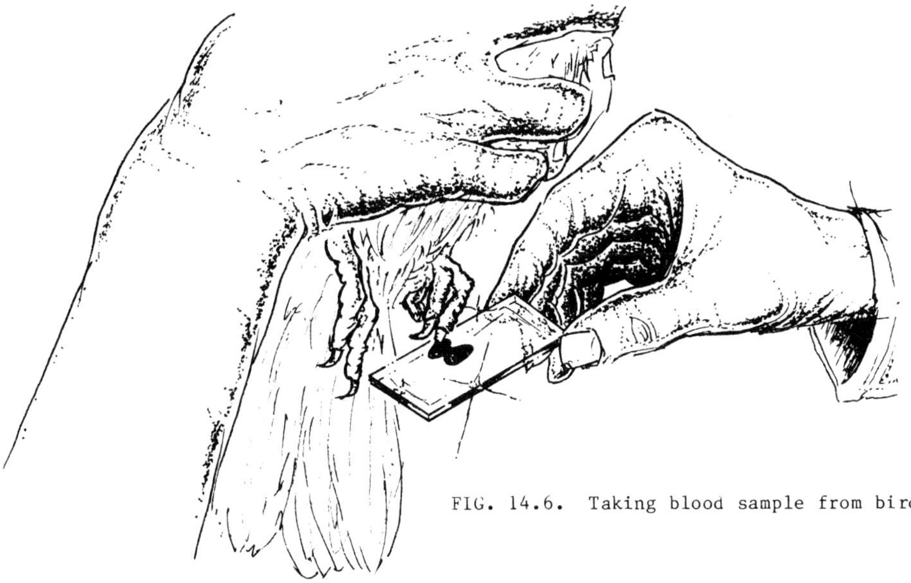

FIG. 14.6. Taking blood sample from bird for blood smear.

Then touch your microscope slide to the toe so that a drop of blood is placed near one end of the slide (Fig. 14.6). Place your second microscope slide on the clean end of the blood-containing slide at a 30° angle. Slide it toward the drop of blood until the blood is touched. Upon contact, the blood will flow to both sides of the second slide (Fig. 14.7). Then push the second slide back toward the clean end of the first slide with uniform pressure. The blood should spread in a uniform smear. Set the slide aside to air-dry thoroughly. See Exp. 15 for staining procedure.

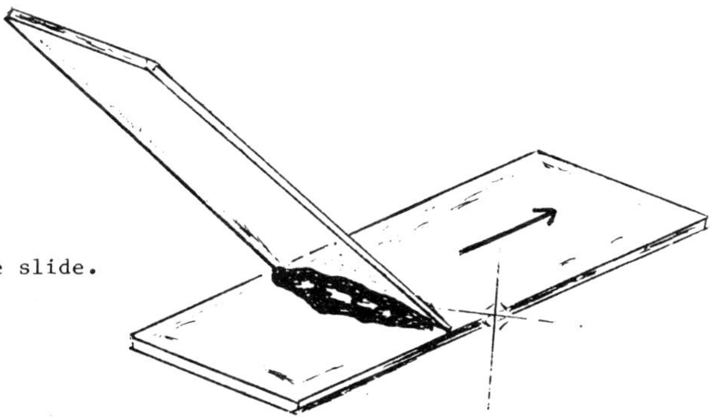

FIG. 14.7. Making blood smear on microscope slide.

If the bird's nail continues to bleed, apply firm pressure with your fingers just dorsally to the cut nail to stop the flow of blood. If it doesn't stop in a minute or so, wipe the toenail clean of blood and cauterize it with a silver nitrate applicator stick. After the toe has stopped bleeding, be sure to rinse off the silver nitrate on the toe by holding the toe under cold tap water, so the bird doesn't pick at the silver nitrate with its beak and burn its tongue.

Reverse roles with your lab partner and repeat the procedure on another toenail.

Brachial vein technique. If a larger volume of blood is needed, a venipuncture can be performed using the brachial vein in the wing. Have your lab partner hold the bird so that its feet are facing the ceiling (dorsal recumbency) and then have your lab partner stretch out one wing.

The brachial vein is located on the ventral surface of the wing (Fig. 14.8). Moisten a cotton ball with 70% alcohol and wipe the feathers in the direction that they lay until they are wet; this makes the brachial vein visible.

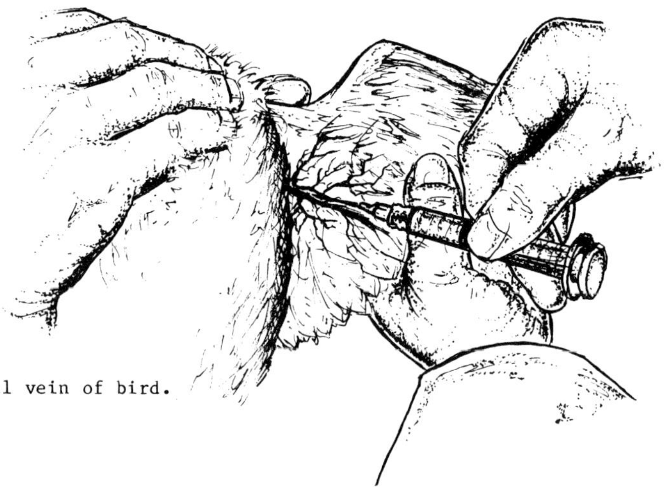

FIG. 14.8. Taking blood sample from brachial vein of bird.

Place your left finger on the vein between the venipuncture site and the bird's heart so that blood will back up in the vein, making it easier to see and puncture. Using the 23 gauge needle and 1 cc syringe, slide the needle into the vein. Exert slight negative pressure by pulling back on the plunger. Blood should flow readily into the syringe. Take only 1/2 cc blood.

To withdraw the needle, release the vein above the injection site, place your finger directly over the injection site and needle, and apply slight pressure as you withdraw the needle. As birds' veins are very fragile and bleed easily, your lab partner must apply pressure for at least 3 min to avoid a hematoma (bleeding under the skin). This is why it is best to take blood from a clipped toenail if possible.

Remove the needle from the syringe and gently squeeze the blood out of the syringe into a centrifuge tube and place it in the centrifuge rack or 100 cc beaker. This blood may be used for additional analysis.

Reverse roles with your lab partner and repeat the procedure with the other wing.

IM Injection

Work in pairs, one partner restraining the bird while the other prepares the bird and gives the injection.

Have your lab partner restrain the bird in dorsal recumbency. Fill the 1 cc syringe with 0.1 cc sterile saline and lay it aside.

Then moisten a cotton ball with 70% alcohol and wet the feathers of the left side of the keel bone (sternum) in the chest area so that the pectoral muscle is visible. This will cause the feathers to lay aside so the injection can be given. Once the alcohol evaporates, the feathers return to normal.

Position the bird so that its head is facing you. Pick up the loaded syringe in your right hand and feel for the keel bone with your left index finger. Insert the needle about 1/4 in. to the left of the keel bone at a 45°

angle (Fig. 14.9) about 1/4 in. into the muscle. Inject the saline. Place your left finger over the injection site and needle and apply slight pressure. With your finger in place, remove the needle. Continue to apply pressure at the site for about 1 min to prevent hemorrhage.

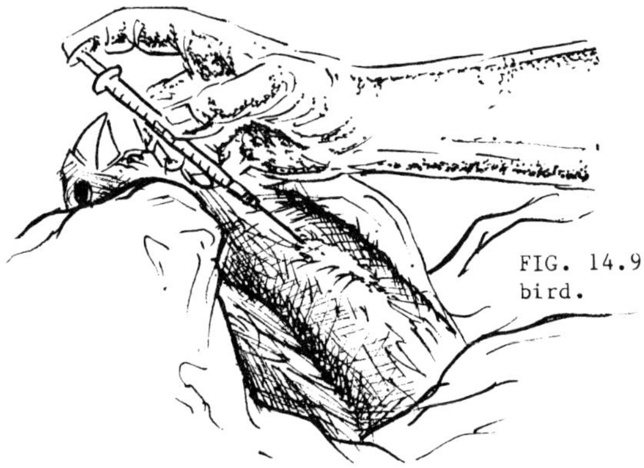

FIG. 14.9. Giving IM injection into pectoral muscle of bird.

Reverse roles with your lab partner and repeat this procedure on the other side of the bird.

Oral Administration

The dose is 0.5 cc/kg tap water. On your student report sheet (number 2), calculate the volume to give your bird.

Have your lab partner restrain the bird in a vertical position. Place the 18 gauge feeding needle on the 1 cc syringe. Fill the syringe with the calculated volume of tap water.

One method of opening a bird's beak is to use your hands. Use your left hand to force the beak open (Fig. 14.10).

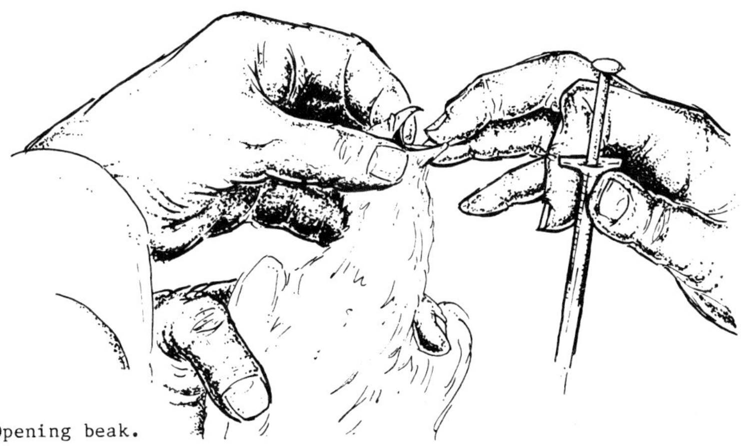

FIG. 14.10. Opening beak.

Or, if the bird is very small, use the forceps method. Pick up the forceps in your left hand, compress the ends together, and slide the ends perpendicularly through the beak. Release the pressure on the forceps so that the ends open, forcing the beak open.

With the feeding needle and syringe in your right hand, allow the tip of the feeding needle to follow the roof of the bird's mouth and pharynx into the crop. Have your partner gently stretch the bird's neck to help facilitate passage of the needle into the crop (Fig. 14.11). Inject the solution slowly and remove the feeding needle.

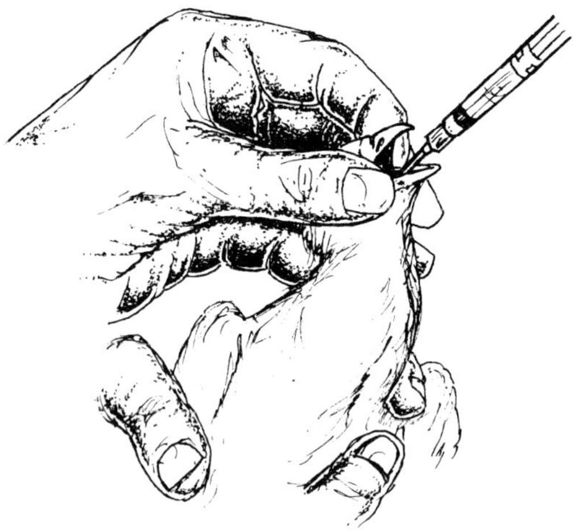

FIG. 14.11. Inserting feeding needle into crop.

Then reverse roles with your lab partner and repeat this procedure.

Nail Trimming

Caged birds usually must have their nails trimmed periodically. Even if the bird is kept on a sandpaper perch, the tips of the nails do not wear down and must be trimmed.

Have your partner restrain the bird. Grasp the bird's foot with your left hand and pick up the toenail clippers in the right hand. Slide the toenail clippers over the nail to a point below (distal to) the quick (end of blood vessel and nerve). If you cannot see the end of the blood vessel, clip the least amount of toenail necessary. If the nail has a split or hangnail, you may file the area with an emery board.

If bleeding occurs, you've cut too much off the nail. Apply direct pressure to the toe above (proximal to) the nail to stop the flow of blood and cut less from the next nail. You may also cauterize the toenail with a silver nitrate stick, but remember to rinse the toenail under tap water so the bird will not burn its tongue on any remaining silver nitrate.

Trim all the nails on one foot. Then reverse roles with your lab partner and trim the nails on the other foot.

Complete your student report sheet.

EXP. 14: STUDENT REPORT

Name______________________________

Date______________________________

1. Weight of bird_________________

2. Does it hurt a bird to have its wings trimmed? Why?

3. The dose of water to be given is 0.5 cc/kg. For a bird weighing 600 g, the volume would be 0.3 cc. Calculate the volume to give your bird (show your work).

4. Give two reasons why a towel might be thrown over a bird before it is picked up.

5. Where should a bird's toenail be clipped for a blood sample?

6. If you use a silver nitrate stick to stop the toenail from bleeding, what must you do before releasing the bird?

7. After you have removed the needle from the venipuncture site, how long must you exert direct pressure on the site? Why?

EXPERIMENT 15

Human, Canine, and Avian Hematocrits and Blood Smears

INTRODUCTION

This experiment acquaints the student with two techniques used to evaluate blood--hematocrits and blood smears.

Blood is composed of two fractions, plasma (the fluid portion) and blood cells (the formed elements). The blood cells can be further divided into different types: red blood cells, white blood cells, and thrombocytes. The red blood cells (RBC, or erythrocytes) contain hemoglobin, which transports oxygen and carbon dioxide. The white blood cells (WBC, or leukocytes) are necessary for the body's defense mechanism. Thrombocytes (blood platelets) are one component necessary for blood to coagulate.

In this experiment, the student collects human, canine, and avian blood and makes a hematocrit determination (percentage of red blood cells) of each. Then the student makes blood smears of human, canine, and avian blood. These smears are stained and the cell types identified using a microscope.

MATERIALS AND EQUIPMENT (per student)

Hematocrit Determinations (human, canine, avian)

- alcohol swabs
- blood lancet
- microhematocrit tube, heparinized
- sealing clay (Seal-Ease Tube Sealer and Holder) (per 4 students)
- microhematocrit centrifuge (per 20 students)
- vacutainer, 5 ml
- needle holder unit for 5 ml vacutainer, disposable, needle size 22 g, 1 in.
- 2 gauze squares, 2 in. x 2 in.
- silver nitrate applicator sticks (1 package per 20 students)
- dog
- bird
- another species of lab animal (optional)

Blood Smears

- alcohol swabs
- blood lancet
- 6 microscope slides
- nail clippers, animal (per 2 students)
- 2 gauze squares, 2 in. x 2 in.
- silver nitrate applicator sticks (1 package per 20 students)

Diff-Quik Stain Set (per 10 students)
3 Coplin jars (per 10 students)
beaker with 500 cc distilled or deionized water (per 10 students)
50 cc beaker (per 10 students)
immersion oil (1 bottle per 20 students)
microscope

PROCEDURE

Hematocrit Determination

A hematocrit determination measures the percentage of packed red blood cells (RBC) contained in the total volume of blood sample. This percentage is found by filling a capillary microhematocrit tube with a blood sample and centrifuging the sample to spin down the formed elements. The height of the RBC column is then compared with the height of the total volume.

Human technique. To make a human hematocrit determination, first take a sample of your or your lab partner's blood. Clean the tip of the left middle or ring finger with an alcohol swab. Place the cleaned finger and hand on the work bench with the palm up. Pick up the lancet in your right hand and stab your finger (or your lab partner's) in the "meaty" end. A drop or two of blood should flow out.

Fill the capillary tube to the correct volume and seal the end you were holding in the sealing clay (Fig. 15.1). (See Exp. 14 to review this technique.)

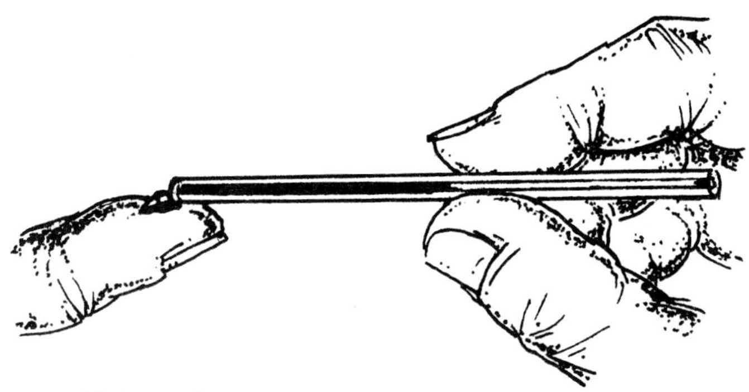

FIG. 15.1. Taking blood sample from finger.

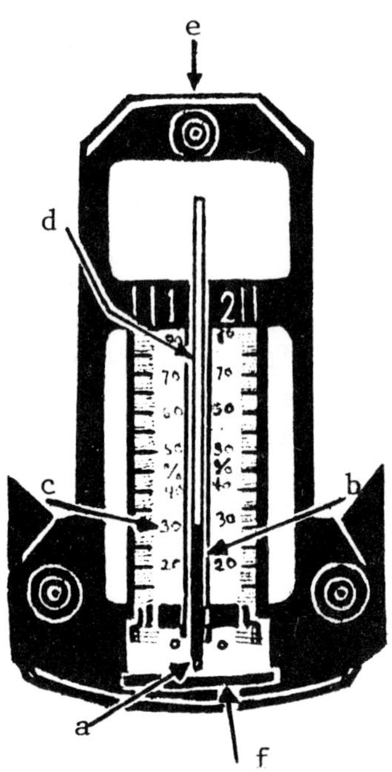

FIG. 15.2. Microhematocrit tube placed in one chamber of centrifuge head: a. clay plug in tube; b. column of red blood cells; c. hematocrit value (30%); d. blood plasma; e. center of centrifuge head; f. rubber cushion of centrifuge head.

Place the capillary tube with the end containing the sealing clay toward the rubber cushion of the centrifuge head (Fig. 15.2). The numbered slot used for each blood sample must be recorded so the tubes can be identified later. Slide the sealed end against the rubber gasket. To balance the centrifuge head, place another blood sample or an empty tube in the slot directly opposite each sample.

After the tubes are loaded, secure the head cover tightly, close the lid, and latch it carefully.

Run the centrifuge for the length of time prescribed by the manufacturer for maximum packing.

After centrifugation, measure and record the total height of all the blood components together and the height of the RBC column (Fig. 15.3).

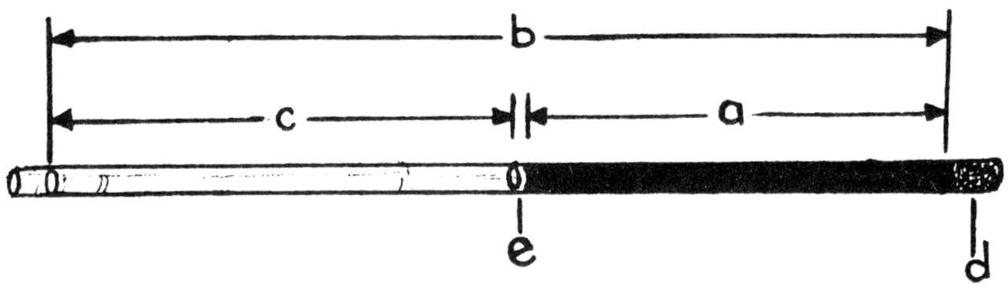

FIG. 15.3. Measuring hematocrit sample: a. column of red blood cells; b. total blood column; c. blood plasma; d. plug in tube; e. buffy coat (leukocytes and thrombocytes).

To calculate the hematocrit, use the equation (a/b) x 100 = h, where a = RBC height, b = total height of all blood components, and h = hematocrit (percentage of packed red blood cells). For example, if a = 40 mm and b = 120 mm, h = (40/120) x 100, or 33%. Or if you are using a centrifuge with a hematocrit scale, slide the hematocrit tube so the 0% mark lies between the sealing clay in the tube and the end of the packed RBC. Then read and record the hematocrit value opposite the junction of the buffy coat and packed RBC.

Canine and avian techniques. Hematocrits are determined by the same method for animals and birds. Take a fresh sample of blood either directly into the capillary tube (mouse's toenail or tail, bird's toenail, rat's toenail or tail, guinea pig's toenail, etc.) or indirectly into a vacutainer. To take blood via a vacutainer, insert the needle into a vein in the same manner as was done for IV administration (for dogs, see Exp. 8; for birds, see Exp. 14). Blood will flow directly into the vacutainer. The blood in the vacutainer is then transferred into a microhematocrit tube. Remove the vacutainer needle and adapter. Tilt the vacutainer so the collected blood flows to the mouth of the vacutainer. Dip the microhematocrit tube into the blood sample (Fig. 15.4). Fill the microhematocrit tube by lowering the end of the tube (see Exp. 14). Determine and record the hematocrit values.

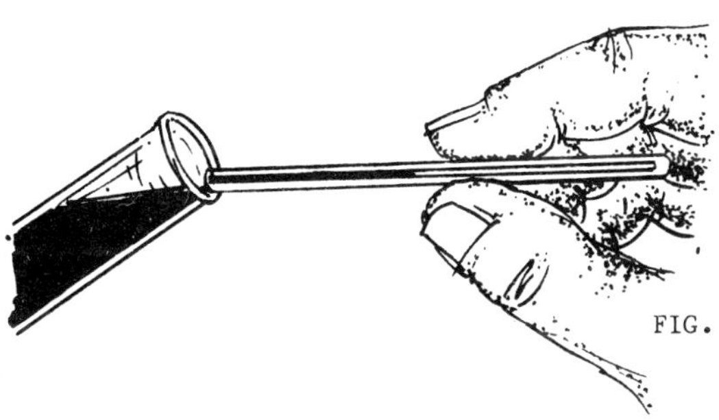

FIG. 15.4. Filling hematocrit tube from vacutainer.

Following are normal hematocrit values for various animals:

Species	Average	Range
cat	40	25-52
dog (beagle)	46	38-53
gerbil	47	41-51
guinea pig	42	37-47
hamster	49	39-59
human	43	40-45
monkey	42	32-52
pigeon	47	42-52
rabbit (New Zealand)	42	33-50
rat	46	39-53

Blood Smears

Blood collection. To make a human blood smear, clean a finger with alcohol, allow it to dry, and prick it with the lancet as was done earlier in this experiment. Allow a drop of blood to fall on one end of the slide. Smear the blood, using another slide (to review the technique, see Exp. 14) and allow the smear to air-dry.

Make a canine blood smear. For a small sample, clip the dog's toenail just above (dorsal to) the quick so the toenail will bleed. Make your smear. To stop the bleeding, use the technique described in Exp. 11A.

Make an avian blood smear from your bird. (See Exp. 14 for the avian blood collection technique.)

Cell types. After your blood smears have air-dried, stain the slides using the Diff-Quik Stain Set (see Exp. 12 for the staining technique.)

After the stained blood smears have air-dried, place each slide on the microscope, add one drop of immersion oil to the smear where the smear appears thinnest, and observe under the oil immersion objective.

To identify the different types of cells on a slide, scan the slide using the Meander system (Fig. 15.5). Be sure to examine the edges of the smears as well as the center. The following is an outline of human, canine, and avian blood cell types. Study your smears and identify these types of cells. Count 100 leukocytes in each smear and record the numbers and types on the student report (numbers 6, 7, and 8). (Your percentages may not be exactly like those in the outline; relative numbers of leukocytes change with disease and individual variation).

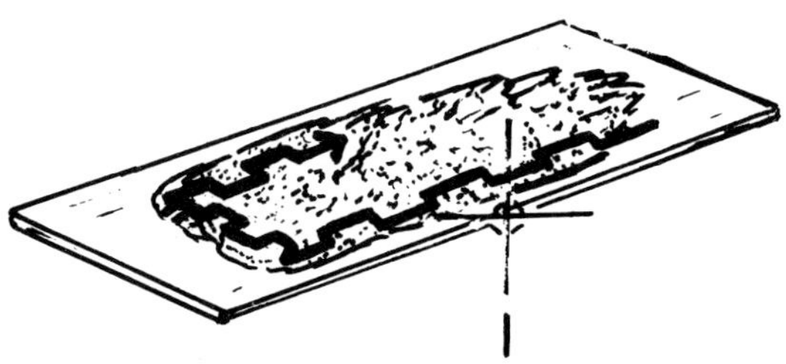

FIG. 15.5. Scanning slide using Meander system.

I. Human blood cells (Fig. 15.6)
   A. Red blood cells (erythrocytes): most numerous, small, round, biconcave discs, pink, without nucleus
   B. White blood cells (leukocytes): 2 main types, further subdivided according to presence or absence of granules in the cytoplasm

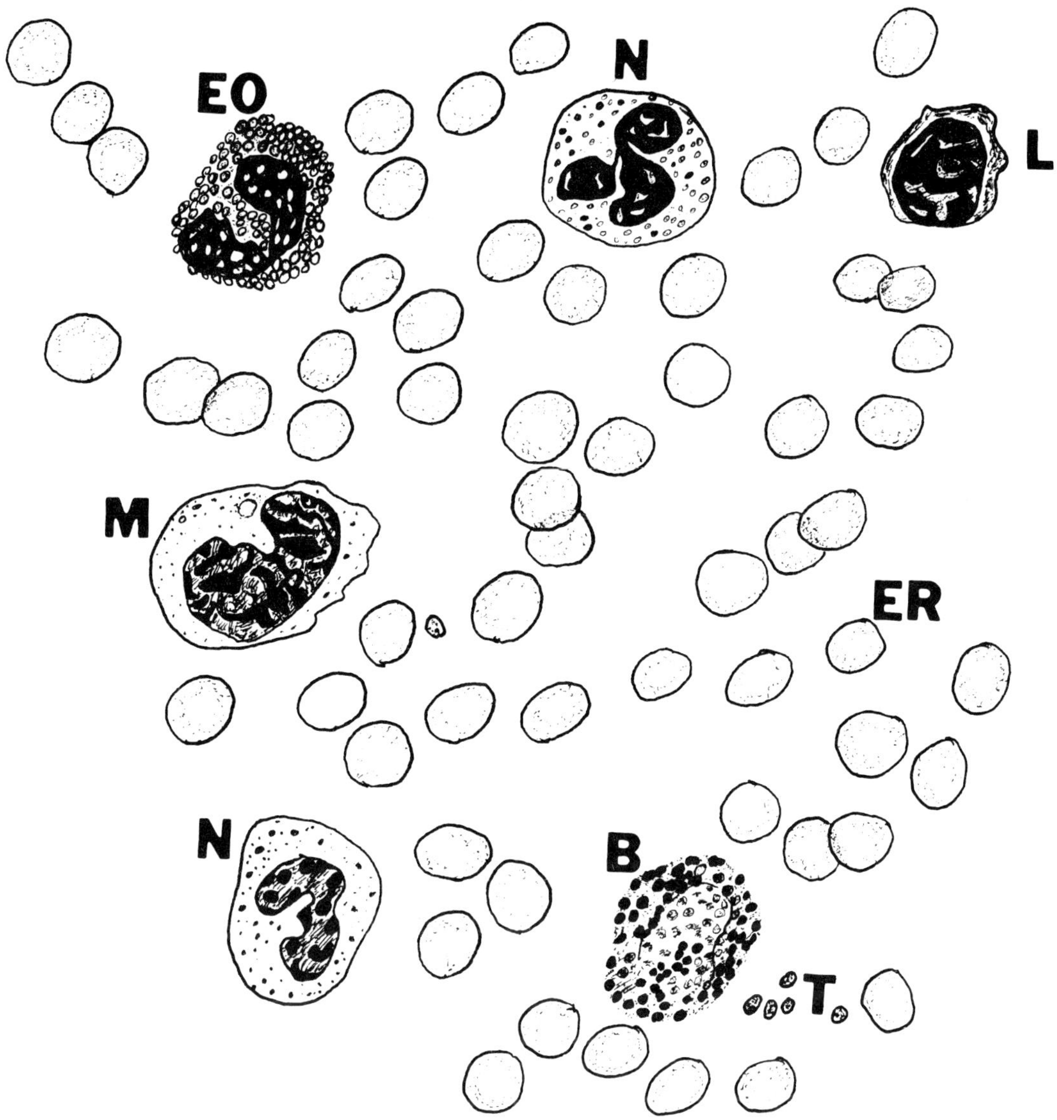

FIG. 15.6. Types of human blood cells: basophil (B); eosinophil (EO); erythrocyte (ER); lymphocyte (L); monocyte (M); neutrophil (N); thrombocyte (T).

1. Polymorphonuclear granulocytes: numerous granules and a lobulated nucleus; 3 types
   a. Neutrophils: lobed, dark blue nucleus; reddish lilac granules in cytoplasm; pale pink cytoplasm; about 65% of total leukocytes
   b. Eosinophils: lobed, blue nucleus; many relatively large, spherical, red-orange granules of equal size in cytoplasm; blue cytoplasm; about 2-4% of total leukocytes
   c. Basophils: round, indented, or bilobed, dark blue nucleus; very dark, coarse, blue granules of unequal size in cytoplasm, usually visible above and below the nucleus; about 0.5% of total leukocytes

2. Agranular leukocytes: few or no granules in cytoplasm, round or horseshoe-shaped nuclei; further subdivided into 2 types
   a. Lymphocytes: size variable, from smaller than RBC to almost twice as large; violet, darker nucleus almost fills cell, usually round or slightly indented; nuclear chromatin clumped and stains darker; cytoplasm a lighter blue color, may contain a few nonspecific darker granules in wider part of the cytoplasm; small leukocytes about 20-25% of total leukocytes, large leukocytes about 3%
   b. Monocytes: generally largest of all cells; large, violet nucleus may be round, lobed, or horseshoe-shaped and not as large as lymphocyte nucleus; chromatin in nucleus arranged linearly rather than arrangement of clumps giving a ground glass appearance; vacuoles may also be present; cytoplasm a sky blue color and may contain fine granules

C. Thrombocytes: smallest of all the formed elements; very small, irregular masses that contain purple granules; no nuclei; usually found together in clusters; 3-7% of total leukocytes

II. Canine blood cells
   A. Red blood cells: generally similar to human
   B. White blood cells: 5 types
      1. Neutrophils: lobed, dark blue nucleus; reddish lilac granules in cytoplasm; pale pink cytoplasm; about 65% of total leukocytes
      2. Eosinophils: generally similar to human granules; vary in size and number; appear vacuolated
      3. Basophils: rarely found in peripheral blood; generally similar to human
      4. Lymphocytes: usually of small type; eccentrically located nucleus
      5. Monocytes: generally similar to human; vacuoles frequently seen in cytoplasm, occasionally seen in nucleus; nucleus extremely variable in shape, usually ameboid
   C. Thrombocytes: similar to human

III. Avian blood cells (Fig. 15.7)
   A. Red blood cells: oval (not round as in human); pink cytoplasm; oval nucleus (no nucleus in human); stains blue; most numerous of all cells
   B. White blood cells: 2 main types, with subdivisions
      1. Polymorphonuclear granulocytes
         a. Heterophils (equivalent of neutrophils in human blood): round; cytoplasm contains numerous elongated bodies called rods that are red-orange in color; some rods may contain granules or vacuoles; nucleus stains purple, is lobed, may not appear connected
         b. Eosinophils: similar to human; differentiated from heterophils by light-blue-staining cytoplasm; red-orange granules of different sizes in cytoplasm; nucleus also lobed but generally stains deeper purple
         c. Basophils: similar to human; nucleus may be masked by very dark blue granules in cytoplasm but sometimes can be seen as lighter-blue-staining body in center of cell; nucleus may appear single-lobed or occasionally bilobed

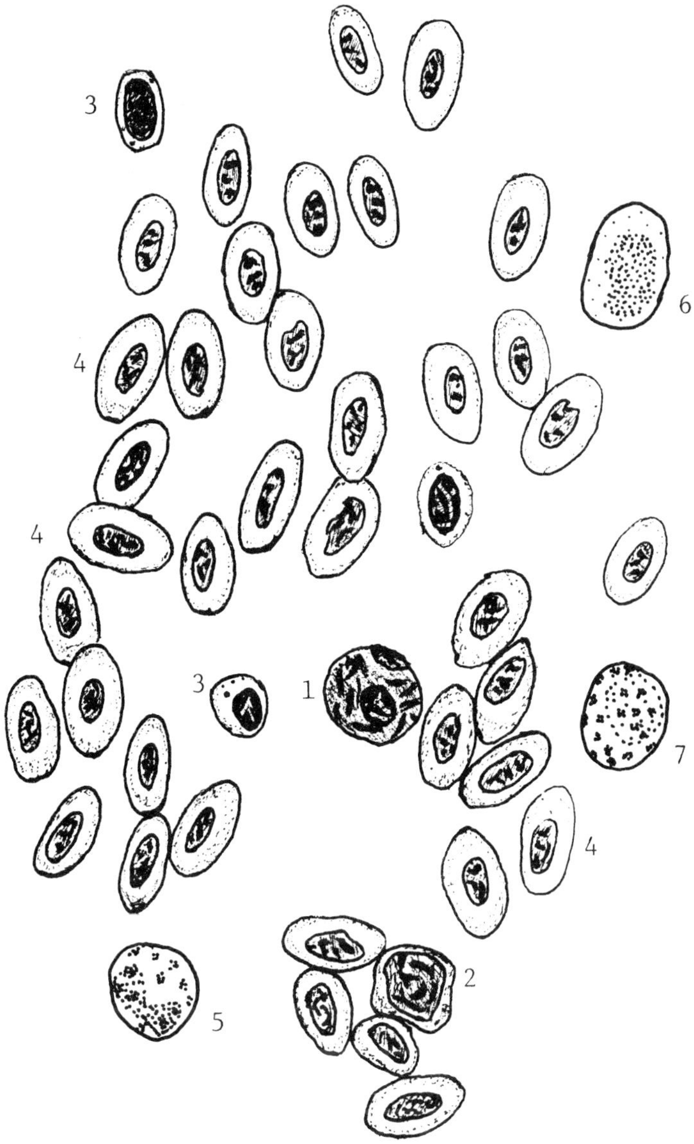

FIG. 15.7. Type of avian blood cells: heterophil (1), lymphocyte (2), thrombocyte (3), erythrocyte (4), eosinophil (5), monocyte (6), basophil (7).

2. Agranular leukocytes
   a. Lymphocytes: similar to human
   b. Monocytes: similar to human; generally slightly larger than lymphocytes; cytoplasm may contain vacuoles (Hof areas) that may stain either very light blue or orange; nuclei sometimes have thinner chromatin clumps than those of lymphocyte

C. Thrombocytes: different from human; mononucleated cell; similar to avian red blood cells but slightly smaller, with rounder ends than red blood cells; usually without the regular oval contour of red blood cells; nucleus usually more rounded, stains darker than red blood cells; cytoplasm stains very light blue and contains one or more darker-staining granules

EXP. 15: STUDENT REPORT

Name________________________

Date________________________

1. Human hematocrit determination:

Height of total blood column in hematocrit tube	__________
Height of red blood cells in hematocrit tube	__________
Hematocrit	__________%

2. Canine hematocrit determination:

Height of total blood column in hematocrit tube	__________
Height of red blood cells in hematocrit tube	__________
Hematocrit	__________%

3. Avian hematocrit determination:

Height of total blood column in hematocrit tube	__________
Height of red blood cells in hematocrit tube	__________
Hematocrit	__________%

4. How might anemia of an individual affect the hematocrit determination and why?

5. What is polycythemia, and how would polycythemia affect the hematocrit determination?

6. Human differential leukocyte count; count 100 leukocytes in the smear and record the numbers of the different types of leukocytes found:

Neutrophils	__________
Eosinophils	__________
Basophils	__________
Lymphocytes	__________
Monocytes	__________
Total	100 cells

7. Dog differential leukocyte count; count 100 leukocytes in the smear and record the numbers of the different types of leukocytes found:

Neutrophils	______
Eosinophils	______
Basophils	______
Lymphocytes	______
Monocytes	______
Total	100 cells

8. Bird differential leukocyte count; count 100 leukocytes in the smear and record the numbers of the different types of leukocytes found:

Heterophils	______
Eosinophils	______
Basophils	______
Lymphocytes	______
Monocytes	______
Total	100 cells

9. If the total number of leukocytes were normally elevated, what condition(s) might be indicated?

10. If the total number of leukocytes were abnormally low, what condition(s) might be indicated?

Appendixes

APPENDIX A. Equipment List

For company addresses, see Appendix C.

Item	Source
Alcohol, ethyl, 95% or 35%	Fisher Scientific Co.
Alcohol, isopropyl, 70%	A. J. Buck & Son
Balance, Mettler or equivalent, 1.000 g	Fisher Scientific Co.
Balance weights: 20 g, 100 g, 1 kg	Fisher Scientific Co.
Bandage, gauze, 2 in. roll	A. J. Buck & Son
Bandage, Conform stretch tape	A. J. Buck & Son
Beakers: 10 cc, 25 cc, 50 cc, 100 cc, 500 cc	Fisher Scientific Co.
Blood lancets	Fisher Scientific Co.
Cages: mouse, rat	Fisher Scientific Co.
Centrifuge, microhematocrit	A. J. Buck & Son
Centrifuge tube, tapered, 13 x 100 mm	Fisher Scientific Co.
Clippers, surgical, small animal	A. J. Buck & Son
Coplin staining jars	Fisher Scientific Co.
Cylinders, graduated: 10 cc, 20 cc, 50 cc, 100 cc	Fisher Scientific Co.
Diff-Quik Stain Set	A. J. Buck & Son
Ether, ethyl (anesthesia grade)	Fisher Scientific Co.
Ether jar (large 1 gal jar with lid, 1 in. of cotton on bottom)	Fisher Scientific Co. or local cafeteria (large mayonnaise or pickle jar)
File metal, triangular	Fisher Scientific Co.
Flashlight, pencil	A. J. Buck & Son
Forceps, tissue, with teeth: 4 in., 5 1/2 in.	A. J. Buck & Son
Funnel, glass, 3-4 in.	Fisher Scientific Co.
Gloves, examination, sterile	A. J. Buck & Son
Gloves, surgical, sterile	A. J. Buck & Son
Hemostats	A. J. Buck & Son
Immersion oil	Fisher Scientific Co.
Lab coats	Fisher Scientific Co.
Marker, indelible (or leather punch)	Local hardware store
Medicine dropper	Fisher Scientific Co.
Microhematocrit tubes, heparinized	A. J. Buck & Son
Microscope with oil immersion lens	Fisher Scientific Co.
Mortar and pestle, 2 1/2 in.	Fisher Scientific Co.

Nail clippers, animal	A. J. Buck & Son
Needle holder	A. J. Buck & Son
Needles, feeding: 16 gauge, 3 in.; 18 gauge, 1 1/2 and 3 in.	Popper & Sons, Inc.
Needles, hypodermic: 20 gauge, 1 and 1/2 in.; 22 gauge, 1/2 in.; 23 gauge, 1 and 1/4 in.; 25 gauge, 1/4, 1/2, and 5/8 in.	A. J. Buck & Son
Needles, suture, heavy cutting, 1/2 circle, C-40 or similar	A. J. Buck & Son
Olive oil, laboratory grade (must be sterilized)	Fisher Scientific Co.
Pads, gauze, 2 x 2 in.	A. J. Buck & Son
Pet piller	A. J. Buck & Son
Restrainer, mouse	Fisher Scientific Co.
Scale, bird (animal balance)	Frey Scientific Co.
Scale, human	A. J. Buck & Son
Scale, mouse	Taconic Farms
Scale, rabbit or baby	A. J. Buck & Son
Scale, rat	Taconic Farms
Scalpel with blade, size 20 blade	A. J. Buck & Son
Scissors, operating, blunt and blunt, 5 1/2 in.	A. J. Buck & Son
Scissors, operating, blunt and sharp, 4 in.	A. J. Buck & Son
Scissors, stitch (suture removal)	A. J. Buck & Son
Sealing clay tube sealer and holder (Seal-Ease Tube Sealer and Holder)	A. J. Buck & Son
Silver nitrate applicator sticks	A. J. Buck & Son
Slides, microscope	Fisher Scientific Co.
Spatula, micro	Fisher Scientific Co.
Stethoscope	A. J. Buck & Son
Stockade, rabbit	Wahmann Manufacturing Co.
Suture, surgical nylon, size 00	A. J. Buck & Son
Swabs, alcohol	Fisher Scientific Co.
Swabs, cotton-tipped applicator	Fisher Scientific Co.
Syringes, plastic, disposable: 1 cc, 3 cc, 10 cc, 20 cc	A. J. Buck & Son
Thermometer, rectal	A. J. Buck & Son
Vacutainer, 5 ml	A. J. Buck & Son
Vacutainer needle holder unit for 5 ml vacutainer	A. J. Buck & Son
Weighing paper	Fisher Scientific Co.

APPENDIX B: Drug List

For company addresses, see Appendix C.

Drug	Supplier
Acepromazine maleate, 25 mg/tablet	A. J. Buck & Son
Alcohol, isopropanol, 70%	A. J. Buck & Son
Alcohol, ethyl, 95% or 35%	Fisher Scientific Co.
Caffeine and sodium benzoate injection, USP, 250 mg/cc	Eli Lilly & Co.
Cerumite, 1/2 fl oz	A. J. Buck & Son
Dextroamphetamine (Dexedrine), 5 mg tablets	Local pharmacy
Doxapram hydrochloride (Dopram-V) injectable, 20 mg/ml	A. J. Buck & Son
Estradiol cypionate (ECP), 2 mg/ml	A. J. Buck & Son
Ketamine hydrochloride (Ketaset), 100 mg/ml	A. J. Buck & Son
Mineral oil, light	Local pharmacy
Pentobarbital sodium (Nembutal), 60 mg/ml	Abbott Laboratories
Petroleum jelly, white (Vaseline)	A. J. Buck & Son
Piperacetazine hydrochloride tablets, 1 mg	A. J. Buck & Son
Saline, normal, sterile, 250 ml	A. J. Buck & Son
Testosterone suspension, 100 mg/ml	Local pharmacy
Thiamylal sodium (Bio-tal, Surital), 1 g	A. J. Buck & Son
Tubocurarine chloride injection, USP	Eli Lilly & Co.
Xylazine (Rompun), 20 mg/cc	A. J. Buck & Son

APPENDIX C: Suppliers

Abbott Laboratories
P.O. Box 68 AP5
Dept. 208
North Chicago, IL 60064

A. J. Buck & Son, Inc.
10534 York Road
Cockeysville, MD 21030

Fisher Scientific Co.
585 Alpha Drive
Pittsburgh, PA 15238

Frey Scientific Co.
905 Hickory Lane
Mansfield, OH 44905

Eli Lilly & Co.
Indianapolis, IN 46285

Popper & Sons, Inc.
300 Denton Ave.
New Hyde Park, NY 11040

Taconic Farms
Germantown, NY 12526

Wahmann Manufacturing Co.
10 West Aylesbury Rd.
Timonium, MD 21093

APPENDIX D: Observation Techniques

A brief description of the observations to be made and technique(s) of determining each is given below.

1. ACTIVITY INCREASE (increase in normal action or motion). Watch for more movement in the treated group than in the control group.
2. ACTIVITY DECREASE (decrease in normal action or motion). Watch for less movement in the treated group than in the control group.
3. ATAXIA (lack of muscular coordination). A staggering gait is obvious in the treated animals.
4. COMA (an unconscious state from which one cannot be aroused). Several or more treated animals should be lying in the cage. Touch them to see if they can be aroused. If touching them elicits no effect, see if they respond to a pain stimulus (see Pain reflex). If the pain stimulus elicits no effect, the animal should be considered to be in a coma.
5. CONVULSION(S) (an intermittent series of involuntary, violent, muscular, rhythmic contractions). One or more animals have uncontrolled muscular contractions (they may be lying on their side and moving their legs or may be jumping abnormally, violently in an uncontrolled manner).
6. CORNEAL REFLEX LOSS (no closing of the eyelid when the cornea is stimulated). The cornea can be stimulated in a larger lab animal (dog, cat, rabbit, etc.) by gently touching the cornea with a wisp of cotton. This reflex is usually not checked in small lab animals such as mice and rats.
7. CYANOSIS (a bluish coloring of the skin and mucous membranes). In the smaller lab animals or white animals this is easily noted by a blue tinge to the ears and tail. In larger lab animals or nonwhite animals the mucous membranes of the mouth or the conjunctiva of the eye should be checked.
8. DEFECATION, EXCESSIVE (diarrhea). Diarrhea is noted easily on the cage bedding or sometimes even on the animal itself.
9. HEART RATE INCREASE (an increase in the number of ventricular contractions of the heart). The beating of the heart can sometimes be observed in the upper abdomen/chest area of small animals that are not moving. Be careful not to mistake respirations for heart rate. In larger lab animals a stethoscope can be used.
10. HEART RATE DECREASE (a decrease in the number of ventricular contractions of the heart). See Heart rate increase.
11. IRRITABILITY INCREASE (abnormal sensitivity to a stimulus). Two different stimuli can be used to elicit the irritability response. The first is another animal in the group accidentally touching the irritable animal. You might observe the irritable animal vocalizing, jumping away, and/or fighting with the animal that touched it. The second stimulus is a tool, such as a pencil. Touch several animals with the eraser end. Observe for abnormal vocalization, agitation, or elusive action.
12. IRRITABILITY DECREASE (abnormal decreased sensitivity to a stimulus). In animals that do not react to the touch stimulus (see Irritability increase), the stimulus for pain (see Pain reflex) can also be used as a measure for decreased irritability.

13. LACRIMATION INCREASE (the increased production and secretion of tears). Tearing can be observed on the cheek of the animal if there is an increase.
14. MUSCLE TONE DECREASE (decrease in the tension of a muscle). This can be tested in a small lab animal by holding it by the middle of its tail or back quarters, letting it grasp a grid (the top of a cage), and then pulling it away from the grid. Observe and compare grip strength to that of control animals. Or if the animal is abnormally inactive, squeeze a large muscle group (upper leg or abdomen muscles) gently to see if the muscle lacks tension.
15. PAIN REFLEX LOSS (loss of reaction to a painful stimulus). To elicit this reaction in small lab animals their tails or footpads may be pinched gently. Observe if the animal reacts by pulling away its foot or tail and compare with the control animals. In larger lab animals such as dogs or cats, pinching a footpad would cause a normal animal to pull the foot away from the stimulus.
16. PALPEBRAL OPENING INCREASE (increase in the opening between the eyelids). Watch for the eyes appearing to pop out of the head.
17. PALPEBRAL OPENING DECREASE (decrease in the opening between the eyelids). Watch for the eyelids closing to some degree.
18. PILOERECTION (erection of hair). Watch for the hair of the animal standing on end. Mice and rats will look like puffballs. The hair of dogs and cats will stand up on the back or neck area as if they were fighting.
19. RESPIRATORY RATE INCREASE (an increase in the number of inspirations per minute). This can be observed in animals that are not active. Observe the chest area, count the number of inspirations, and compare with the control animals.
20. RESPIRATORY RATE DECREASE (a decrease in the number of inspirations per minute). See Respiratory rate increase.
21. RIGHTING REFLEX LOSS (the inability of an individual to right itself after being placed on its side). Place the animal on its side, if possible, and see if the animal has enough coordination to get to its feet.
22. SALIVATION INCREASE (increase in the production of saliva). Excessive salivation will be obvious on the chin and neck areas of small lab animals. Larger lab animals will drool.
23. STARTLE RESPONSE INCREASE (an increase in the response to being startled or alarmed unexpectedly). Run a pencil along the metal cage top or side to create a sudden loud noise. Watch for animals jumping abnormally, compared with the reaction of the control group.
24. STARTLE RESPONSE DECREASE (a decrease in the normal startle response). See Startle response increase.
25. STRAUB RESPONSE (a response that causes an animal to hold its tail erect in the air). Watch for the treated animals carrying their tails higher than the control animals.
26. TOUCH RESPONSE INCREASE (an increase in response to an unexpected touch). The animal will jump when touched.
27. TOUCH RESPONSE DECREASE (a decrease in response to an unexpected touch). The animal will not react or will react with less intensity than normal when touched.
28. TREMORS, TWITCHES, JERKS (involuntary trembling, twitching, and/or jerking of muscles). Watch for trembling, twitching, or jerking.
29. URINATION INCREASE (increase in the excretion of urine). Observe the bedding or paper under the animal's cage for increased wetness.

30. VASOCONSTRICTION (constriction or decrease in the diameter of blood vessels). The extremities (especially the ears and tail) or the mucous membranes of the animal appear less pink than those of normal animals.
31. VASODILATION (an increase in the diameter of the blood vessels). The extremities and mucous membranes of the animal appear abnormally pink or red.
32. VOCALIZATION INCREASE (an increase in sound production). The treated animals will make more noise than the controls.

Glossary

ACTIVITY--action or motion
AGRANULAR--having no granules
AMPULE--small, hermetically sealed glass container containing medication
ANALGESIA--absence of sensibility to pain without loss of consciousness
ANEMIA--reduction below normal of volume or number of red blood cells or in quantity of hemoglobin in the blood
ANESTHESIA--loss of feeling or sensation
ANTAGONIST--drug that binds to a cellular receptor instead of a hormone, neurotransmitter, or another drug, blocking the action of that substance without producing any other physiological effect
ANTERIOR--pertaining to or situated toward the front
ASEPTIC--preventing infection
ATAXIA--total or partial lack of muscular coordination
ATRIUM--one of a pair of chambers of the upper heart
AUDITORY MEATUS--passage in ear leading to eardrum
AVIAN--pertaining to birds
BASOPHIL--granular leukocyte with irregularly shaped nucleus that is partially constricted into two lobes; cytoplasm stains readily with basic dyes and contains coarse granules of variable size
BICONCAVE--having two surfaces that are indented (or depressed) in the middle
BILOBED--having two well-defined portions
BUFFY COAT--portion of hematocrit sample that contains leukocytes and thrombocytes; buff color
CALIBRATION--determination of the accuracy of an instrument to ascertain necessary correction factors
CANINE--pertaining to dogs
CASTRATION--the removal or destruction of gonads
CAUDAL--pertaining to tail or posterior end
CERUMINOUS--pertaining to a waxy secretion in the ear
CHROMATIN--the material of the nucleus that stains with basic dyes
CNS--central nervous system; the portion of the central nervous system consisting of brain and spinal cord
COAGULATORY GLAND--small gland located in the scrotum of the adult male rat
COMA--state of unconsciousness from which individual cannot be aroused
CONTROL ANIMAL--a normal animal used as a standard against which experimental observations can be evaluated

CONVULSION--series of involuntary, generally violent, muscular contractions, rhythmically alternated with periods of relaxation.
CORNEA--clear, transparent anterior portion of eye covering iris and pupil
CORNEAL REFLEX--reflex action resulting in automatic closing of eyelid when cornea is stimulated (cornea can be stimulated by gently touching it with a wisp of cotton)
CORNIFIED--converted to a horny tissue; keratinized
COVERT FEATHER--feather covering the bases of quills of the wings and tail of a bird
CRANIAL--pertaining to the head or anterior end
CROP (of a bird)--pouched enlargement of gullet of many birds that serves as receptacle for food and for preliminary food softening
CYANOSIS--bluish tinge of skin and mucous membranes due to deficient levels of oxygen in blood
CYTOPLASM--protoplasm of cell that surrounds nucleus
DEFECATION--elimination of feces
DEGENERATION (adjective, DEGENERATIVE)--decline of effective power, vitality, or essential quality
DIFFERENTIAL LEUKOCTYE COUNT--determination of the percentage of different types of leukocytes present in a blood smear
DISSOCIATIVE ANESTHESIA--drug that causes individual to be unaware of its environment
DORSAL--pertaining to the back or situated on the back
DORSOVENTRAL--pertaining to the back and belly; lying on the back with the belly up
DOSE--amount of medicine or agent administered at one time to an individual
DRUG--any medical substance used internally or externally for treatment, cure, or prevention of disease
DUCTUS DEFERENS--see VAS DEFERENS
EMESIS--act of vomiting
EOSINOPHIL--granular leukocyte with a nucleus that usually has two lobes and cytoplasm that contains coarse, round granules of uniform size; stains readily by eosin dye
EPITHELIUM (adjective, EPITHELIAL)--covering of external and internal surfaces of the body
ERYTHROCYTE--red blood cell or corpuscle; one of the formed elements of blood
ESOPHAGUS--passageway extending from the pharynx to stomach
ESTRADIOL--estrogenic steroid, isolated from ovaries and urine of pregnant mares and prepared semisynthetically by hydrogenation of estrone
ESTRUS (adjective, ESTROUS)--recurrent period of female sexual receptivity in mammals
EUTHANASIA--easy or painless death
FEBRILE--pertaining to or characterized by fever
GRANULE--small particle or grain
GRANULOCYTE--any cell that contains granules, especially the polymorphonuclear granulocytes of the blood (neutrophils, eosinophils, and basophils)
HEART RATE--number of contractions of cardiac ventricles per minute
HEMATOCRIT--volume percentage of packed red blood cells in whole blood
HEMATOMA--localized collection of extravascular blood, usually clotted in an organ, space, or tissue

HEMOSTAT--instrument used to constrict blood vessel to check the flow of blood

HETEROPHIL--granular polymorphonuclear leukocyte found in birds; similar to neutrophil in man

HORMONE--substance produced by cells and transmitted by blood to other cells and organs on which it has specific regulatory effect

HYPNOSIS--artificially induced passive state that leads to state of light or deep sleep

HYPODERMOCLYSIS--introduction of fluids in large quantities into subcutaneous tissue

INSPIRATION--drawing of air into lungs

INTRAMUSCULAR (IM)--injection directly into muscle

INTRAPERITONEAL (IP)--injection directly into abdominal cavity

INTRAVENOUS (IV)--injection directly into vein

INTUBATION--insertion of a rubber, plastic or metal tube into body organ or duct

IRRITABILITY--ability of organism or tissue to react to stimulus

JERK--sudden reflex or involuntary movement

KEEL BONE--sternum; breastbone

LACRIMATION--secretion and discharge of tears

LATERAL--pertaining to the side

LD_{50}--dose that when given to a group of test animals will cause 50% of the animals to die

LEUKOCYTES--any of white or colorless nucleated cells found in blood; white blood cells

LIGATURE--any suture material, such as thread or wire, that is used to tie tissues or organs during surgery

LOBE--rounded projection

LOBULATED--divided into or made up of lobes

LYMPHOCYTE--type of white blood cell, agranular, having one nucleus, formed by lymphoid tissue, and necessary for immunity

MIDLINE--imaginary line that divides the body into right and left sides

MONOCYTE--large, agranular, phagocytic leukocyte with relatively small, eccentric, oval or kidney-shaped nucleus

MUSCLE TONE--normal degree of vigor and tension of muscle

NARCOTIC--agent that produces stupor or insensibility

NEUTROPHIL--granular leukocyte having a three- to five-lobed nucleus; cytoplasm contains very fine granules and stains readily with neutral dyes

NUCLEUS (plural, NUCLEI; adjective, NUCLEATED)--spheroid body inside cell, contained inside a double membrane, with chromosomes and one or more nucleoli

ORAL--pertaining to mouth

OVARIECTOMY--removal of ovaries

OVULATION--discharge of mature ovum or ova from ovary

PAIN REFLEX--response to painful stimulus that causes an automatic muscular reaction

PALPATE--to feel with the hand

PALPEBRAL OPENING--open area between eyelids

PARAMETER--distinguishing feature that can be measured, either directly or indirectly

PARENTERAL ROUTE--not through alimentary canal (e.g., by way of subcutaneous, intramuscular, intrasternal, or intravenous injection)

PERCENT SOLUTION--grams per 100 cc solution

PERIPHERAL BLOOD--blood located near surface, outside central region
PERITONEUM--serous, shiny membrane lining the walls of pelvic and abdominal cavities and enclosing viscera
PER OS (PO)--by mouth; given through the mouth
PHARYNX--canal connecting the oral cavity and larynx with the esophagus
PILOERECTION--erection of hair
PINNA--cartilaginous, projecting portion of external ear; ear flap
PLASMA--fluid portion of unclotted blood (see also SERUM)
POLYCYTHEMIA--excess number of red corpuscles in the blood
POLYMORPHONUCLEAR--having nucleus deeply lobed or so divided as to appear to have multiple nuclei
PREANESTHESIA--light, preliminary anesthesia induced by medication given prior general anesthesia
PROXIMAL--nearest to point of reference
PRURITIS--itching
PSYCHOTROPIC--pertaining to drug that affects mental state; capable of modifying mental activity
Q.S.--*quantum sufficit*; to add enough liquid to make a certain volume
RECTUM (adjective, RECTAL)--distal portion of large intestine, beginning as continuation of the sigmoid colon and ending at anal canal
RECUMBENT--lying down
RESPIRATION--the act of breathing in and out once
RESPIRATORY RATE--number of inspiratory or expiratory breaths measured per minute
RIGHTING REFLEX--ability to right oneself (stand up) after being laid on side or back
ROTATE--to turn around; to twist
ROUTE--prescribed method of applying or injecting a compound (i.e. orally, topically, subcutaneously)
SALIVATION--secretion of saliva
SCROTUM--pouch that contains testes and their accessory organs in most male mammals
SCRUFF--loose skin on animal's back
SEMINAL VESICLES--pouches on either side of male reproductive tract that temporarily store semen
SERUM--clear fluid that separates from blood upon clotting; plasma from which fibrinogen has been removed by clotting process
SOLUTION--preparation of one or more soluble substances dissolved in liquid medium
SPERMATIC VEIN--vein that drains blood from testes and epididymis
STARTLE RESPONSE--response to being frightened or alarmed unexpectedly
STERNUM--breastbone
STRAUB RESPONSE--response that causes animal to hold its tail erect
STRIATED MUSCLE--skeletal muscle controlled by conscious part of brain
SUBCUTANEOUS (SC)--injection between skin and muscle layer
SUSPENSION--preparation of finely ground, undissolved substance dispersed in liquid medium
SYSTOLE--contraction, or period of contraction, of the heart, especially that of the ventricles
TARE--to weigh a container prior to weighing a substance together with container
TESTES--male gonads; egg-shaped glands that produce spermatozoa and testosterone, situated normally in scrotum

TESTOSTERONE--male hormone produced by testes that is responsible for inducing and maintaining male secondary sex characteristics
THROMBOCYTE--blood platelet
TO EFFECT--(to give medication) until an obvious effect is produced
TOPICAL--local application applied externally
TOUCH RESPONSE--response elicited when touched unexpectedly
TRANQUILIZER--agent that calms or quiets an anxious individual
TREMORS--involuntary trembling of body or limbs
TUNICA VAGINALIS--serous membrane covering the sides and front of the testes and epididymis
TWITCH--brief contraction of muscle
UMBILICUS--the cicatrix (new tissue formed) marking the site of the attachment of the umbilical cord in the fetus
UREA--compound found in urine, blood, and lymph, formed from amino acids and ammonia compounds
URINATION--act of excreting urine
VACUOLATED--pertaining to or characterized by vacuoles
VACUOLE--small space or cavity formed in the protoplasm of a cell
VAGINAL SMEAR--specimen of vaginal cells that is examined microscopically to determine the estrous or menstrual stage of a female
VAS DEFERENS--ductus deferens; excretory duct of the testis, which unites with the excretory duct of the seminal vesicle
VASOCONSTRICTION--constriction, or decrease in size, of blood vessels
VASODILATION--dilation, or enlargement, of blood vessels
VEHICLE--substance in which drug or compound is dissolved, suspended, or mixed
VENTRAL--pertaining to abdomen; directed toward belly surface
VENTRICLE--one of the two lower chambers of heart
VIAL--small glass container or bottle
VITAL SIGNS--signs of life, usually pulse, respiration, and temperature
WINK REFLEX--reflex causing eyelid to close when median corner of eye is touched

Bibliography

Angel, J. E., ed. 1984. Physicians' Desk Reference. 38th ed. Oradell, N.J.: Medical Economics.

Aronson, C. E., ed. 1983. Veterinary Pharmaceuticals and Biologicals 1982/83. Edwardsville, Kans: Veterinary Medicine Publishing.

Arrington, L. R. 1978. Introductory Laboratory Animal Science. Danville, Ill.: Interstate Printers and Publishers.

Barnes, C. D., and L. G. Eltherington. 1965. Drug Dosage in Laboratory Animals: A Handbook. Berkeley: University of California Press.

Booth, N. H., and L. E. McDonald, eds. 1982. Veterinary Pharmacology and Therapeutics. Ames: Iowa State University Press.

Catcott, E. J., ed. 1977. Animal Health Technology. 2d ed. Santa Barbara, Calif.: American Veterinary Publications.

Coles, E. H. 1980. Veterinary Clinical Pathology. Philadelphia: W. B. Saunders.

D'Amour, F. E., F. R. Blood, and D. A. Belden, Jr. 1965. Manual for Laboratory Work in Mammalian Physiology. Chicago: University of Chicago Press.

Diggs, L. W., D. Sturm, and A. Bell. 1954. The Morphology of Blood Cells. North Chicago: Abbott Laboratories.

Fowler, M. E. 1978. Restraint and Handling of Wild and Domestic Animals. Ames: Iowa State University Press.

Giovanoni, R., and R. G. Warren. 1983. Principles of Pharmacology. St. Louis: C. V. Mosby.

Goodman, L. S., and A. G. Gilman. 1980. The Pharmacological Basis of Therapeutics. 6th ed. New York: Macmillan.

Hafez, E. S. E., ed. 1970. Reproduction and Breeding Techniques for Laboratory Animals. Philadelphia: Lea and Febiger.

Harkness, J. E., and J. E. Wagner. 1983. The Biology and Medicine of Rabbits and Rodents. 2d ed. Philadelphia: Lea and Febiger.

Inglis, J. E. 1980. Introduction to Laboratory Animal Science and Technology. Elmsford, N.Y.: Pergamon.

Kirk, R. L., ed. 1980. Current Veterinary Therapy: VII. Small Animal Practice. Philadelphia: W. B. Saunders.

Lane, D. R., ed. 1980. Jones's Animal Nursing. Elmsford, N.Y.: Pergamon.

Lucas, A. M., and C. Jamroz. 1961. Atlas of Avian Hematology. Washington, D.C.: USDA

Petrak, M. L. 1982. Diseases of Cage and Aviary Birds. Philadelphia: Lea and Febiger.

Schalm, O. W., N. C. Jain, and E. J. Carroll. 1975. Veterinary Hematology. 3d ed. Philadelphia: Lea and Febiger.

Shirley, B. A. 1975. Laboratory Manual of Mammalian Physiology. New York: Macmillan.

Siegmund, O. H., ed. 1979. The Merck Veterinary Manual. 5th ed. Rahway, N.J.: Merck.

Sturkie, P. D. 1976. Avian Physiology. 3rd ed. New York: Springer-Verlag.

Thompson, T., and C. R. Schuster. 1968. Behavioral Pharmacology. Englewood Cliffs, N.J.: Prentice-Hall.

Upson, D. W. 1981. Clinical Veterinary Pharmacology. Bonner Springs, Kans.: V. M. Publishing.

Walker, W. F., Jr. 1970. Dissection of the Rat. San Francisco: W. H. Freeman.

Waynforth, H. B. 1980. Experimental and Surgical Technique in the Rat. New York: Academic.

Williams, C. S. F. 1976. Practical Guide to Laboratory Animals. St. Louis: C. V. Mosby.

Woerpel, R. W., and W. J. Rosskopf. 1981. Avian Handling Techniques: Part I. Animal Health Technician 2 (6):325-31.

Woerpel, R. W., and W. J. Rosskopf. 1982. Avian Handling Techniques: Part II. Animal Health Technician 3 (2):81-85.

Index